The Talking Telephone

and 14 Other Custom Telephone Projects

Steve Sokolowski

TAB

TAB BOOKS

Blue Ridge Summit, PA

FIRST EDITION
FIRST PRINTING

© 1990 by TAB BOOKS
TAB BOOKS is a divison of McGraw-Hill, Inc.

Printed in the United States of America. All Rights Reserved. The publisher takes no
responsibility for the use of any of the materials or methods described in this book,
or for the products thereof.

Library of Congress Cataloging-in-Publication Data

Sokolowski, Steve.
    The talking telephone—and 14 other custom telephone projects /
by Steve Sokolowski.
       p.   cm.
    ISBN 0-8306-7571-X   ISBN 0-8306-3571-8 (pbk.)
    1. Telephone—Equipment and supplies—Design and construction—
Amateurs' manuals. I. Title.
TK9951.S66   1990   90-36056
621.385—dc20   CIP

TAB BOOKS offers software for sale. For information and a catalog, please contact
TAB Software Department, Blue Ridge Summit, PA 17294-0850.

Questions regarding the content of this book
should be addressed to:

    Reader Inquiry Branch
    TAB BOOKS
    Blue Ridge Summit, PA 17294-0214

Acquisitions Editor: Roland S. Phelps
Book Editor: B.J. Peterson
Production: Katherine Brown
Cover Design: Lori E. Schlosser

Contents

Acknowledgments

I wish to extend my heartfelt thanks to the following companies and corporations for supplying me the needed technical information as well as the photographs and illustrations on their products.

Allen-Bradley Co.
Augat
Cortelco/ITT Phone System
Del-Phone Industries
Modern Electronics
Radio Electronics
Teltone Corporation

Introduction

In the beginning, Bell invented the telephone. Over the years, developments in technology provided high-tech communication devices that have very little resemblance to the original, 100-year-old discovery. Developments like speaker phones, electronic ringers, music-on-hold, digital data transmissions, and the like can trace their conception at some point in history.

The advent of telecommunications has come a long way since the now-famous words ''Mr. Watson, Come here. I want you.'' were first transmitted over copper wires just a little over a century ago. Electronic technology has provided us, the children of the twenty-first century, miraculous innovations that, in the words of some obscure person, will make our lives easier. Contrary to this belief, our lives have been transformed into an unbearingly awkward and hard-to-manage 9-to-5 routine. This drudgery is the result of ever-increasing amounts of information gushing out of computer terminals all over the world.

The technology to tame the 1990s is here now and so are the hefty prices that usually accompany newly developed ideas. To domesticate the fears of hobbyists and electronic technicians alike, this book was written. The purpose of this book is to illustrate simple and economical ways to enhance the operation of any rotary or tone-dial telephone. This is done by using reliable but inexpensive as well as innovative integrated circuits and project designs.

In the pages of this book, I will take you on a fascinating adventure through the world of telecommunications, digital electronics, speech synthesizers, and control circuitry. But all journeys, just like the one we are about to take, begin with the first step.

So there we will start.

Advanced telephone theory is discussed. In addition, a chapter is presented to acquaint the hobbyist with the tools and equipment needed to assembly and intelligently troubleshoot any electronic project. Whether it's a telephone enhancement circuit or a more complicated computer layout, you will be armed with the best possible tools of the trade.

Upon completion of these two chapters, our telcom adventure really begins. I will show you how to take a handful of easily obtainable parts and convert them into an inexpensive telephone ringer that rivals, in performance, the more complicated and expensive commercial products. Then just by adding another few integrated circuits, you can magically transform the telephone ringer into a conversational English style ringer with its distinctive two-ring audio output.

You also will find ingenious ways to assemble circuits that will provide a voice for your telephone. One cleverly designed circuit will actually allow your telephone to speak, in a clear voice, the telephone number as you dial. A circuit of this type not only provides an audio verification of the number before it is processed by the central office, it also supplies the needed help that a visually impaired person needs to make correctly dialed telephone calls. I am proud to play a part in the design and distribution of such a profoundly indispensable tool such as this.

Our adventure is not over yet. Not only can you make your telephone speak a number as it is being dialed, you can also provide an easy way of allowing the same phone to cry out ''alert'' or something like ''attention'' every time a call is being received. This project eliminates the need for the old-fashioned bell-type ringer and replaces it with a twenty-first century work of art.

A talking telephone is not the only ingenious project you will find in this book. I will also show you how to design telecommunications projects using digital circuits, telcom integrated circuits and memory chips. Also I have included schematics to build three working programmable read-only memory programmers. These programmers are a must for any budding telcom engineer. And best of all, I've made arrangements with Del-Phone Industries to provide you with the highly sophisticated integrated circuits and other related components that you will not find in your local electronics store. They include the dual tone multi frequency receivers, dial-pulse counters, crystals, the printed circuit board for the Talking Telephone, as well as other projects. For the address of Del-Phone Industries, please see appendix D. Send them a self-addressed, stamped envelope and they will be happy to send you a list of available components and current prices.

For this book, I could go on and on talking about the highly-sophisticated circuits and projects you will find within these pages. The only way you can appreciate the time and effort that a number of engineers from Del-Phone Industries have spent in the development of many of the projects is for you to build a few.

Telephone Basics

THE TELEPHONE HAS BEEN THE MOST WIDELY USED PIECE OF ELECTRONIC APPARATUS ever introduced to the public. Although the appearance of Alexander Graham Bell's original invention is nowhere near its present day configuration, this telecommunication marvel has allowed hundreds of millions of people to communicate over the entire globe just by entering a predetermined number (the *telephone number*) into sophisticated electronic switching equipment. And all can be accomplished by lifting a piece of plastic (the *telephone handset*), bringing it to your ear, listening for a tone to be generated at the receiver (the *dial tone*), then selecting the predetermined number by either *pulsing* (using a rotary dial) or *tone generation* (Touch Tone dial). The two instruments can then be connected through the miracle of either mechanical or electronic switching.

It staggers the mind to think that any two telephones, located hundreds or even thousands of miles apart, can be linked to allow the human voice or computer data to pass effortlessly between them. Yet this massive electronic network can be accessed and controlled by an unskilled operator: you.

Telephones are as common as leaves on a tree, but their internal operation is known to a relatively few. It is the purpose of this book to bridge this gap with informative discussions, innovative circuits, and practical ideas—ideas of which can be the building blocks of knowledge where your own designs can be nurtured and shaped into practical new telephone accessories.

THE JOURNEY BEGINS

The longest journey begins with the first step. So here, the discussion of the hows and whys of telephone communication begins.

The telephone itself can be classified into two main categories: the *tone dial* (Touch Tone) and the older, *rotary dial* telephones. Two additional subcategories can be added to the list. They are: *business telephone systems* (multiline) and the *home telephone* (single line). Seeing that home telephones are the main concern of this book, the category of business telephones is placed (excuse the pun) on hold for the time being.

The most common of all telephones is the standard, single-line, desk rotary phone (type 500). Except for the updated plug-in coil cord on the telephone handle, this telephone has really never had any major changes since the 1950s.

A more updated instrument is the single-line, desk, tone-dial telephone (type 2500). This telephone also is available as a wall-mounted unit (type 2554). The 2554 telephone has the same basic internal components as the 500 and 2500 except that the 2554 makes use of a single-gong ringer (type 148BA). The 500 and 2500 use the larger, two-gong ringer (type 130BA).

The Trendline rotary desk telephone. Unlike all the previous telephones discussed, the Trendline (type 200) contains a rotary dial in the handle for easier dialing. If you dial a wrong number, you can press the small button located just below the rotary dial. This button disconnects the telephone from the line for a moment or two. When you release the button, the dial tone will reappear. At this time, you can dial again. The Trendline telephone also is available in Touch Tone as well as wall models.

These are just a few of the telephone designs that are common in a home. But since the court ordered breakup of AT&T, *decor* (decorative) telephones have flooded the market. These telephones can be in just about any shape or size that the imaginations of design engineers can come up with. These phones can take the shape of Walt Disney, cartoon, and television characters. Telephones in the shape of a duck are quickly becoming a common sight in major department stores. What would Bell say if he knew how today's society redesigned his brain child?

Whatever the shape the telephone might come in, it still has the same basic parts and the same basic operation. It is these similarities that you will examine next.

THE HANDSET

Whichever telephone you own, whether a standard desk or a newer decor phone, there must be some device that allows you to listen and talk comfortably to someone on the other end. The device in question is called the *telephone* handset. Figure 1-1 illustrates the components and structure of a typical handset. More than just a device that will allow the operator to make use of the instrument easily, the handset is also a housing, usually made of plastic. Located within the housing are the two main operating components: the *receiver* and *transmitter*.

Transmitter

It is the function of the transmitter to transform the variations in air pressure that results from the movement of sound waves. The sound waves in question are in the form of the spoken word. Figure 1-2 depicts the internal structure of the carbon transmitter. As seen in this illustration, sound waves bombard a thin skin called the *diaphragm*. This bombardment causes the diaphragm to vibrate in proportion to the amount of pressure that is exerted on it. The diaphragm, in turn, is physically connected to a reservoir of carbon granules. By compressing and expanding the granules, a resistance that varies in

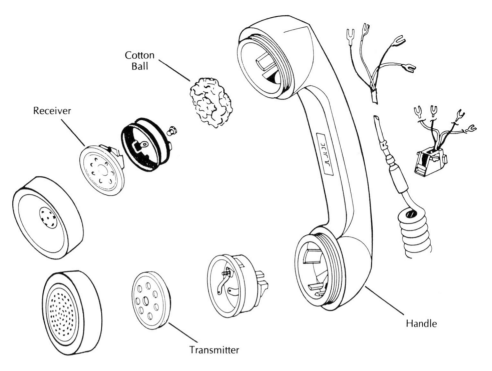

Cotton
Ball

Receiver

Transmitter

Handle

Fig. 1-1   *An exploded view of a standard telephone handset
(Courtesy of Cortelco/ITT Phone Systems).*

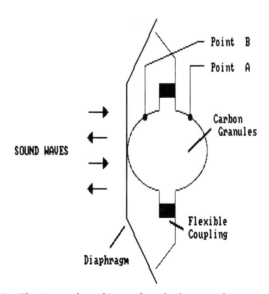

Point B

Point A

Carbon
Granules

SOUND WAVES

Flexible
Coupling

Diaphragm

Fig. 1-2   *The internal workings of a telephone carbon transmitter.*

proportion to the applied pressure is developed at output points A and B. By applying an electrical voltage (this voltage is called the *talk battery*), a current is produced. It is this current that is passed along the telephone lines just waiting to be transformed back into the original speech pattern by the receiver.

Receiver

The receiver transforms the varying current produced by the transmitter back into an intelligible speech signal. A typical electromagnet receiver is shown in Fig. 1-3. A permanently magnetized soft iron core is encapsulated by many turns of very fine wire. In this wire, the varying electrical current is applied. This current attracts and repels an iron diaphragm located on the front of the receiver. The vibrating action that it creates, produces different variations in air pressure. These differences are translated by the ear into useful speech.

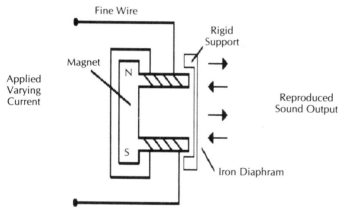

Fig. 1-3   *The internal mechanics of a typical telephone receiver element.*

Figure 1-4 illustrates how an audio signal is converted, by the transmitter, into electrical variations and then back into the original speech pattern by the receiver. Notice the current meter. This meter can be added to produce a visual indication of the varying current. This indication is obtained by allowing a needle (or pointer) to flip vigorously back and forth in tempo with the applied audio. If you wish, you can assemble this demonstration circuit in a few minutes by using an old flea market telephone. Just connect the positive (+) side of a power supply, adjusted to about 6 Vdc (volts direct current), to one terminal of the transmitter. Connect the other terminal of the transmitter to a *dropping resistor* of about 100 Ω (ohms). Complete the circuit by wiring in the current meter (observe polarity) and the receiver. When the circuit is complete, talk into the transmitter. Observe how the pointer of the meter swings wildly. Also note that your voice is also being faithfully reproduced by the receiver (this reproduction is called the *side tone*). By the way, this circuit can be used as a telephone intercom system. Just add longer lengths of wire between elements.

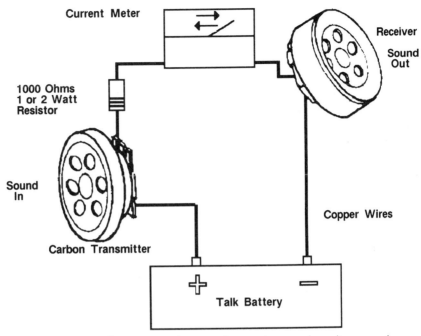

Fig. 1-4 *By taking a transmitter, receiver, resistor, and a voltage supply,*
you have the makings of a simple, yet operational, telephone system.

THE TELEPHONE NETWORK

If you have ever opened a 500 or 2500 telephone, you might have noticed either a metal box with screw mountings on the top or a printed circuit board with push-on terminals. If you have not, now would be a good time to flip to appendix C, where you will find exploded diagrams for the most common home telephones. Included are the 500 and 2500 instruments. Note the un- usual presence of a PC (printed circuit) board. This board (usually found in newer telephones—an electronic box is found in older models) contains the needed electronics for telephone operation. This device, whether it is a PC board or box component, is called the *telephone network.* You might ask: "If Fig. 1-4 illustrated a working telephone, why do you need the network?" Pic- tured in Fig. 1-5 is a 190107 telephone network assembly. It is the function of this network and the PC Board network to provide all the components and *ter- mination points* (either screw or push-on terminals) needed. The components and termination points connect and match the impedance of a typical tele- phone handset transmitter and receiver units to a two-wire telephone circuit.

Both types of telephone networks incorporate RF (radio frequency) filters and side-tone balancing circuits in addition to the impedance-matching com- ponents. As shown in Fig. 1-5, all components for the 190107 network are mounted to the underside of a molded terminal board that is clipped to a mounting container filled with sealing compound. Figure 1-6 illustrates the

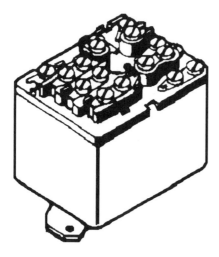

Fig. 1-5 *A typical box type telephone network. These networks can be seen in many older telephone models.*

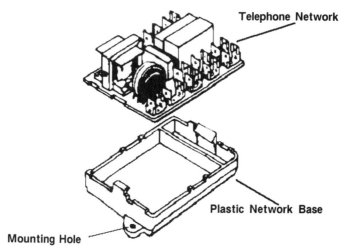

Telephone Network

Plastic Network Base

Mounting Hole

Fig. 1-6 *The newly manufactured telephones make use of a network mounted on a printed circuit board.*

newer printed circuit board network. The physical appearance of the two networks are dramatically different, but both contain basically the same components. Both perform the same function.

The telephone network can be compared to an engine block in a car. Somehow, all telephone components must be connected to the network, just as all engine components must be connected to the block. The network is usually the most reliable component in the phone, especially if the older box network is used. All delicate components are protected by the metal enclosure. In contrast, the PC board network is a very fragile item, especially the matching transformer mounted on it. If you look closely at a board network, you can see that very fine wires are protruding from the component. The wires are soldered to the terminal legs. A careless slip of a screwdriver can spell disaster for the phone. There is no way these broken wires can be repaired. So if you decide to assemble a project in this book, use care if you must remove the cover from any telephone.

THE TELEPHONE HOOK-SWITCH

A means of disconnecting the telephone from the line is needed. The easiest way is to have the unused handset resting in a cradle. While resting, the handset can apply pressure to a spring-loaded operating arm, which is directly connected to a number of switch contacts. You probably know what component this is—it is the telephone hook-switch. As mentioned above, the function of the hook-switch is to disconnect the telephone from the line when a call is completed. And in some telephones (like the 2554) connects the 148BA ringer across the line waiting for the burst of 30 Hz (Hertz) voltage that will ring the phone.

Figure 1-7 illustrates a typical telephone hook-switch and its location under the cradle in a 500 or 2500 telephone. Note how the plunger sits gracefully on the operating arm just waiting for the return of the handset or the pressure exerted by someone's finger.

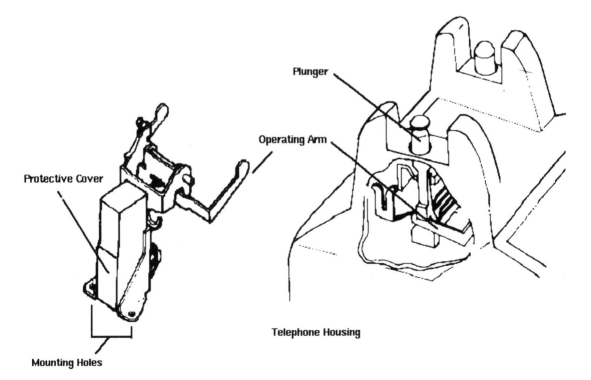

Hook-Switch Assembly

Fig. 1-7 *A telephone hook-switch that is used in a typical single-line desk phone.*

Because of the difference in styling between the wall and desk telephones, wall phones created a problem all their own. Specially designed hook-switches had to be created, too, not only to hang up the phone but to supply a place to store the handset when not in use. Figure 1-8 is only one design that was developed by ITT. This hook-switch operates in the same fashion a 500 or 2500 component does. But this item was designed specifically to fit snugly inside a model 554 or 3554 telephone. Other designs have been developed for the Trendline and the 2554 wall phones. Designs and physical appearances differ, but they all have one purpose in common—to disconnect the phone from the line when not in use.

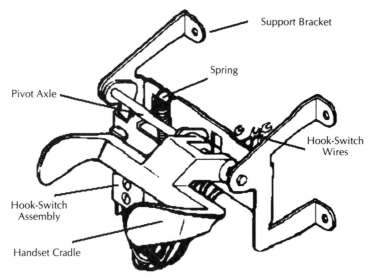

Support Bracket

Spring

Pivot Axle

Hook-Switch Wires

Hook-Switch Assembly

Handset Cradle

Fig. 1-8 *Desk phones make use of one hook-switch style, but a wall telephone has its own design as shown here.*

THE TELEPHONE RINGER

Once a call has been connected, the central office that serves the area of the called party must send out some kind of signal to indicate that a call is waiting. This signaling function is called *ringing*. The type of signal used is an ac (alternating current) voltage in the range of 90 to 110 V with a frequency of 30 Hz. In the earlier days of the telephone, *party lines* were used almost extensively. These installations connected a number of telephones in parallel. Of cause, if two people are talking on the line, a third or even a fourth party can listen in on the conversation. This connection scheme did not provide any privacy at all. But how did the telephone company signal one particular telephone to ring in the midst of five or six units? The answer is by frequency selection. The bells used in the older telephones provided different capacitor and ringer coil impedance values. It was these variations that made the bell select one and only one frequency on which to be activated.

For example, say there are five telephones connected to a party line (telephone 1 through telephone 5). If a call were placed to phone 1, the central office would send a ringing signal of 16 Hz to the common party line. Phone 1 would ring. All the others would remain silent. If a call to phone 5 were made, the CO (central office) will send a ring signal of 66 Hz to the line. At that time, telephone 5 would ring at 66 Hz. All other phones would remain silent. This scheme is something like radio tuning. You turn the knob to the desired broadcast frequency of the station you wish to hear. All other transmissions will experience what is called a high impedance at this tuner location. Thus the radio will not ''see'' or tune in these other radio frequencies.

Figure 1-9A depicts a block diagram of a standard two coil telephone ringer, and B depicts the wiring of a single coil bell. By applying a voltage

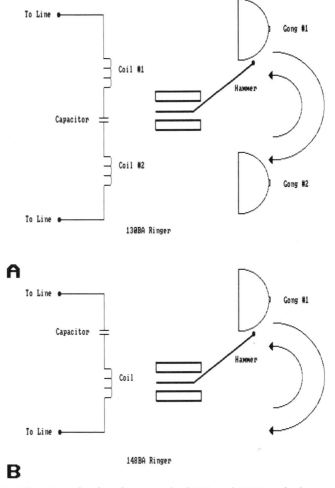

Fig. 1-9 *The wiring hookup for a standard 130 and 148BA telephone ringer.*

where indicated, a series resonant circuit is produced (just like the radio tuner but at a much lower frequency). The 30 Hz signal, which is also the resonant frequency of ringers used today, generate a magnetic field around a device called a *hammer*. This hammer is attracted and repelled by the changing magnetic field generated by the incoming ring signal. If two gongs were placed on either side of the hammer, the hammer would strike each gong as it goes through its flip/flop swing. Thus an audio signal is sounded to indicate an incoming call.

The operation of a single gong ringer, as illustrated in Fig. 1-9B is the same as the two-gong bell, except one gong is used instead of two. This makes the single-gong ringer more compact. This compactness can be seen in Fig. 1-10. This is an illustration of the type 148BA ringer. Due to its condensed size, the 148BA is the perfect choice for the Trendline desk and wall telephones as well as the 2554 wall phone.

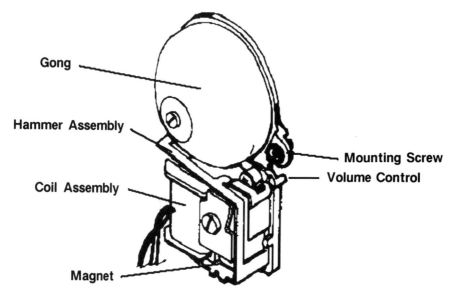

Fig. 1-10 *A typical 148BA telephone ringer. These ringers can be found in the more compact Trendline desk and wall telephones.*

Figure 1-11 depicts the common 130BA ringer. Unlike its counterpart, the 148BA, the 130 is large, bulky and heavy. Ringers of this type can be found in large telephones such as the 500, 2500, and the 3554.

The ring duration is also a standard signal provided by the CO. In the United States, the standard is 2 seconds on and 4 seconds off. In England, the CO provides two short bursts of power with a 2 second rest period between. Figure 1-12 illustrates this difference quite well. Note that all times shown in this illustration are in seconds. Also notice the ring used in Great Britain. One of the projects in this book (the English style telephone ringer) simulates this unique signal. It is a project that truly deserves your time in construction.

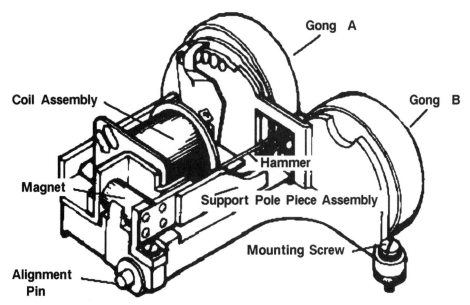

Fig. 1-11 *The 130BA ringer. These ringers are usually found in larger single-line, as well as multiline desk and wall telephones.*

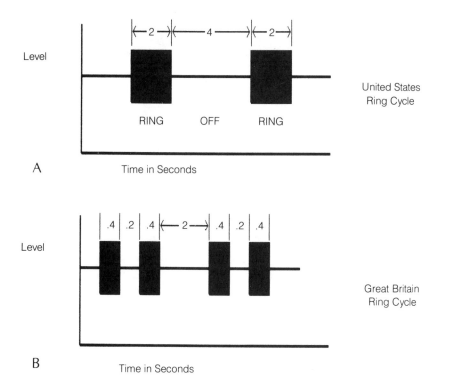

Fig. 1-12 *Ring signal durations (in seconds) for the United States and Great Britain.*

THE ROTARY DIAL

In a conventional telephone like the 500, dialing a number is accomplished by a device called a rotary dial. This item creates equally spaced make-and-break pulses in accordance to how far the plastic dialing plate has been rotated in the clockwise direction. Figure 1-13 is a good illustration of a model 30 rotary dial. Regularly spaced holes in a dialing plate called the *finger wheel* and a metal bracket, called the *finger stop*, make the rotation of the dial to a predetermined number an easy operation. Each hole in the finger wheel represents a single digit. These digits range from the 1 to 10. But of course, the tenth and last hole in the plate is the designation for the operator (OPER or 0).

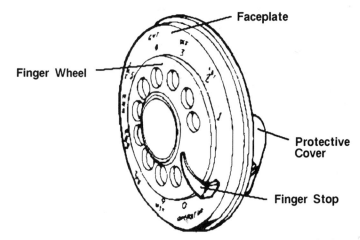

Fig. 1-13   *A 30G rotary dial. Used mainly in the 500 series telephones.*

By using a number of small gears and a device that dictates the speed of the return of the finger wheel to its resting position (this device is actually called the *governor*) internal switches (*pulse contacts*) are opened and closed at a rate of 1 pulse per second with a make-break ratio of 61.5 percent (Fig. 1-14). The number of pulses created by the normally closed contacts is determined by how far the finger wheel was advanced before its movement was ended by the finger stop. In other words, if you place your finger into the fifth hole of the finger wheel and rotate it until you reach the finger stop, the internal contacts would open and then close five times. If you rotate the wheel until the 0 hole reaches the finger stop, there would be an opening and closing of the contacts 10 times.

The string of pulses created by the rotary dial can be seen in Fig. 1-15. This illustration depicts the make-break ratio, the on and off Hook status, and the pulses created by dialing the number 4. Just remember that the pulsating contacts of the dial are in the normally-closed configuration. Pulses are made by the opening, not the closing, of the contacts.

When the finger wheel is moved from its resting position, a second set of contacts close. Unlike the pulsing contacts, this second set (see receiver-shorting contacts in Fig. 1-14) of contacts close and remain closed throughout the

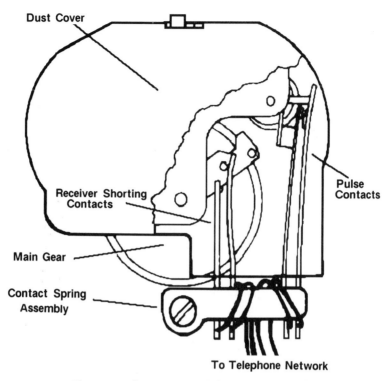

Fig. 1-14 *The rear view of the 30G rotary dial.*

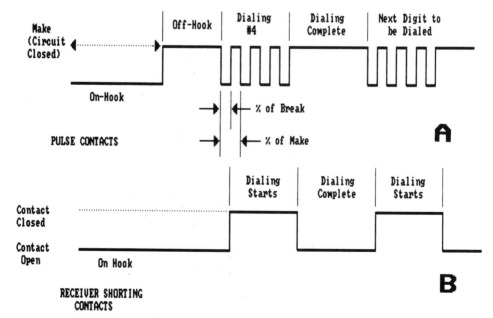

Fig. 1-15 *To determine the desired number being dialed, the central office equipment has strict make and break ratios that every rotary dial must abide by.*

entire *pulse train* (a name associated with the number of make and break pulses created by the dial at any given finger wheel rotation). It is the function of this switch to short the telephone receiver during the dialing period. If this switch were missing, you would hear loud and annoying clicks in the receiver. By shorting the receiver, these clicks are prevented from being heard.

Protruding from the dial are four wires, terminated with spade lugs: 1 green wire, 1 blue wire, and 2 white wires. The green and blue wires are from the dial pulsating contacts and are connected to the telephone network by either screwing them in place or by pushing them into specially designed terminals. The two white wires are from the receiver shorting contacts. They too, are connected to the network.

The rotary dial is mounted by using location studs on either side of the main supporting bracket (Fig. 1-16). These studs slide easily over the left and right dial brackets, which are riveted to the telephone's base plate. Screws are then tightened to secure the dial in place.

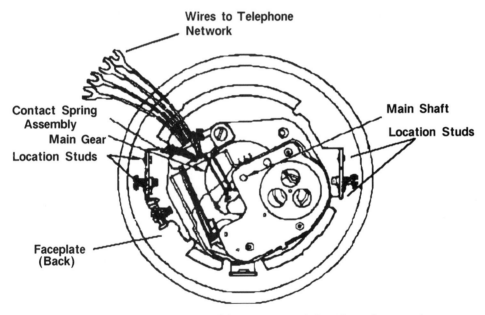

Fig. 1-16 *Another rear view of the 30G rotary dial (without dust cover).*

The procedure that seems most difficult for the novice to learn is: How to remove and replace the dial finger wheel. It is not extremely complicated. But once you learn, it will become second nature. To help understand how to disassemble the finger wheel, read the following step-by-step instructions:

1. Rotate the finger wheel completely clockwise.
2. Insert the straightened end of a paper clip or similar tool into the tab-release hole (Fig. 1-17) that is now approximately 1/4 inch to the left of the tip of the finger.

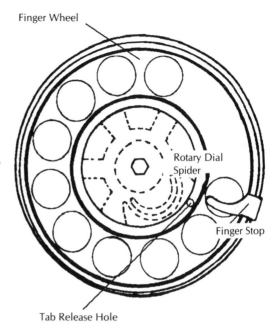

Finger Wheel

Fig. 1-17 *To remove a rotary dial finger wheel, place a paper clip in the tab release, hold and press while turning the wheel past the zero stop.*

Rotary Dial Spider

Finger Stop

Tab Release Hole

3. Press down on the paper clip to spring the tab of the spider. Rotate the finger plate clockwise to release it. Work the finger plate off the spider and out from under the finger stop.

4. Now with the finger wheel removed from the dial, replacement of the telephone number card is easy.

For additional information on the mounting and wiring of a rotary dial into a 500, 3554, or even the Trendline telephone, see appendix C.

TONE DIALING

In addition to the pulsing-out of desired telephone numbers, another and more interesting scheme was developed. Telephones, such as the 2500, 3554, and the 2554 are equipped with a device that creates and impresses audio tones on the telephone line. It is these tones that are transformed by a central office into the desired number being dialed. But of course, the central office must be equipped with the appropriate filtering devices that can be used to detect these specially selected tones.

The act of placing an audio signal on the telephone line as a dialing medium is called DTMF (dual tone multifrequency) dialing. Not to be confused with the normal speech tones also on the line, the tone dial generates a combination of two tones. These tones are created by the high and low group frequencies. When any button is pressed on the dial pad, row and column tones are combined to create a third signal. It is this third tone that is sent down the telephone line and detected at the central office. Once detected, the tone is transformed back into the desired number.

Figure 1-18 shows the relationship between the high and low frequency groups. From this illustration, the buttons 1, 2, 3, and A are assigned the low frequency of 697 Hz. The numbers 4, 5, 6 and B are assigned the frequency 770 Hz. The remaining two rows of buttons also are assigned frequencies as shown. The high group is, in reality, the same as the low, except that the high frequencies are located in columns. As an example, buttons 1, 4, 7 and * generate a tone with the frequency of 1209 Hz. The 2, 5, 8, and the 0 buttons are assigned the frequency 1336 Hz. This scheme also holds true for the remaining two columns. You might have noticed that Fig. 1-18 depicts the addition of a fourth column of buttons that are not included on a standard tone-dial telephone. The column making up the buttons A, B, C, and D are used for special telcom and central office purposes and are not included on a 2500 telephone. But when engineers talk about DTMF signaling, it is assumed that these additional buttons are included in this conversation. For this book, these buttons will not be included in any discussions.

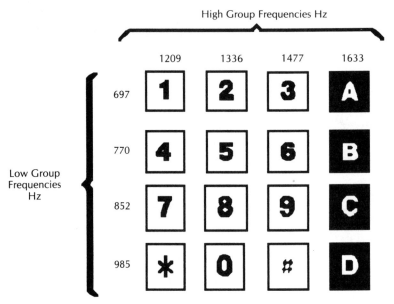

Fig. 1-18 *Tone dials generate two groups of tones. This drawing indicates both the column and row frequencies. Note that buttons A to D are also shown. These buttons are used for special signaling purposes and are not found on a standard home telephone.*

Taking the discussion one step further, there are 12 buttons arranged in a pattern of four rows by three columns. Imagine an oscilloscope connected across the telephone. Press the 7, 8, and 9 buttons on the tone dial. Besides noticing the normal dc voltage (talk battery), you will be confronted with a wave form that is shown in Fig. 1-19. The frequency of this signal, as indicated in Fig. 1-18, is 852 Hz (± 1.5 percent). The other remaining row tones can be seen by just pressing the corresponding buttons. Now, press all the buttons in column 2. The oscilloscope will display a wave form illustrated in Fig. 1-20.

A Row 3, 852Hz. DTMF Signal

Fig. 1-19 *If you would hook up an oscilloscope at the output of a tone dial and press the 7, 8, and 9 buttons simultaneously, the scope would show a sine wave at the frequency of 852 Hz.*

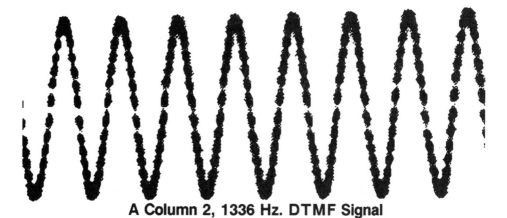

A Column 2, 1336 Hz. DTMF Signal

Fig. 1-20 *If you would press the 2, 5, 8, and 0 buttons simultaneously, you would see this waveform on the scope with a frequency of 1336 Hz.*

This tone has a frequency of 1336 Hz. As with the tones generated by pressing all the buttons in the row group, the remaining high tones can be generated and seen on the oscilloscope, just by pressing the associated buttons.

But what would happen if you just press one button? In keeping with the discussion, press the number 8 button. Figure 1-21 shows that the row tone (852 Hz) and the column tone (1336 Hz) have been internally combined by the tone dial to create a totally new signal. It is the physical characteristics of this third signal that allows the central office to distinguish the difference between the signals generated by the tone dial and the frequencies created by a normal voice pattern. By pressing the remaining eleven buttons, the scope will display the other waveform patterns generated by the dial. With the available twelve distinctive signals, any telephone number can be dialed. By using the * and the # buttons, you can have a circuit detect these buttons, and have it place calls on hold. This project is no fantasy; it is in this book.

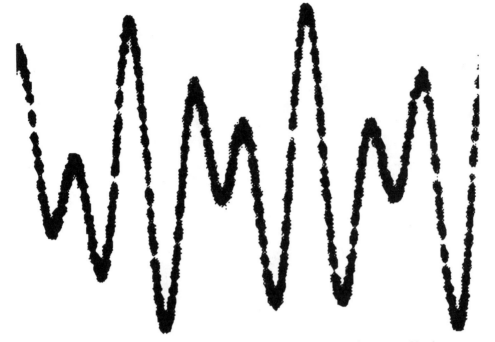

Fig. 1-21 *By pressing the number 8 button, you will see a combination of both tones.*

TONE DIALERS

Through the advancements of technology, Touch Tone dials can be divided into two distinctive categories: the older capacitor/inductor circuit and the newer and more cost effective digital IC (integrated circuit) type. Figure 1-22 shows the older 32G tone dial that is still in use today. Although it uses technology of the 1960s, the 32G is a workhorse. Because of the impeccable workmanship of its assembly, it provides the needed reliability for today's switching equipment.

The 32G also has companion dials. For speaker phone installations, the 36G, which is the same as the 32G except for the addition of switching contacts, is used in its place. For installations requiring nonpolarity sensitive hookups, the 32opg dial is used. Except for the addition of four diodes that make up a bridge rectifier, the 32opg is the same as the 32G.

To illustrate this point, schematics of both the 32 and 36 Touch Tone dials are shown. Figure 1-23 clearly shows the relationship between both dials. Except for the additional switching contacts (Fig. 1-23B—violet and green/white wires), the dials are the same. If a bridge rectifier were installed, opg dials can be made.

Assume you have a 32G tone dial. You can add a bridge. This addition can change the old 32G into a newer 32opg. If you have a 36G dial and add a bridge

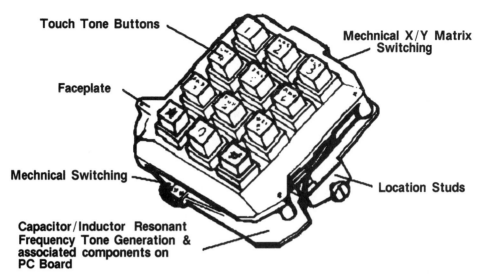

Touch Tone Buttons

Mechnical X/Y Matrix Switching

Faceplate

Mechnical Switching

Location Studs

Capacitor/Inductor Resonant Frequency Tone Generation & associated components on PC Board

Fig. 1-22 *A typical 32G (or 32opg) type Touch Tone dial.*

rectifier, you have a new 36opg dial. Just how to install the needed four diodes to make a telephone independent of line polarity is discussed and illustrated in the telcom cookbook (appendix A).

To continue, turn your attention to the mechanical makeup of the dial. Unlike the newer digital dial pads, the 32 and 36 series of devices require that a series of tiny switches be opened or closed while a button is being pressed. To see how this is done, take a look at Fig. 1-24. A conventional tone pad is made from a faceplate, tone-dial buttons, 12 springs, a slide, seven cranks, and finally a common switch assembly. This seems to be a lot of parts to get a dial to operate, and it is. The illustration shows the faceplate, buttons, and four horizontal and three vertical cranks. The springs have been omitted from this illustration to simplify the drawing.

The cranks are specially designed bars that fit both horizontally and vertically on the back side of the faceplate, over the buttons. These bars are designed in such a way that when one button is pressed, one vertical and one horizontal bar is made to twist on its axis. Note that on the horizontal cranks there are two levers on the left side and two on the right. Also note these levers on the three vertical cranks as well. When the cranks twist by the pressing of a button, these levers move downward, forcing electrical switches located beneath them to close. These switches are located around the pads faceplate, just below all the levers. When a number is dialed, corresponding levers force contact with its associated switches. This scheme is called a mechanical X/Y matrix.

Figure 1-23 shows that these switches are connected to tap-offs on the tuning coils of the dial. Each tap-off is selected to deliver the desired resonant frequencies (Fig. 1-18 frequencies) when activated by the closing of an associated

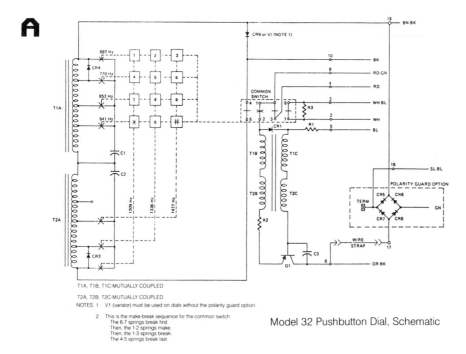

T1A, T1B, T1C-MUTUALLY COUPLED

T2A, T2B, T2C-MUTUALLY COUPLED

NOTES: 1. V1 (varistor) must be used on dials without the polarity guard option.

2. This is the make-break sequence for the common switch.
    The 6-7 springs break first.
    Then, the 1-2 springs make.
    Then, the 1-3 springs break.
    The 4-5 springs break last.

Model 32 Pushbutton Dial, Schematic

Fig. 1-23   *The internal electronics of the 32G tone dial (A) and a 36G speaker phone tone dial (B). (Note the added switching contacts on the common switch.) (Courtesy of Cortelco/ITT Phone Systems.)*

switch. To complete the resonant frequency scheme of the pads, capacitors (C1 and C2 in Fig. 1-23) are placed in parallel with the coils.

In addition to a pair of cranks twisting when buttons are pressed, Fig. 1-25 illustrates a comblike component that slides down when any button is pressed. It is this downward movement that opens or closes contacts that are located on the common switch assembly. The schematic presented in Fig. 1-23 also illustrates the wiring of this all important switch.

Figure 1-26 shows the relative position of the slide and the common switch assembly as seen from the backside of a standard tone dial. Also seen is the mounting of the PC (printed circuit) board. It is this board that contains all the electronics and tuning coils that generate the DTMF signal. Figure 1-27 gives a more detailed drawing of the component locations and the odd shape of the board needed to assemble a compact tone dial. Note in the upper right hand corner of Figure 1-27. These four diodes make up the bridge rectifier. On a 32G, these components will be missing; on a 32opg, they will be included.

DIGITAL TONE DIAL (THE 42opg)

Shown in Fig. 1-21 is the waveform produced by the oscillations of the inductor/capacitor resonant circuit of a 32G tone dial (also the 36G, 32opg, and

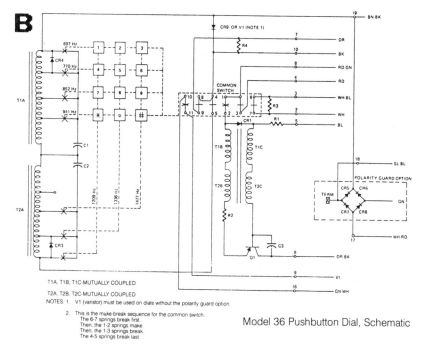

T1A, T1B, T1C-MUTUALLY COUPLED

T2A, T2B, T2C-MUTUALLY COUPLED

NOTES: 1. V1 (varistor) must be used on dials without the polarity guard option.

2. This is the make-break sequence for the common switch.
    The 6-7 springs break first.
    Then, the 1-2 springs make.
    Then, the 1-3 springs break.
    The 4-5 springs break last.

Model 36 Pushbutton Dial, Schematic

Fig. 1-23 *Continued*

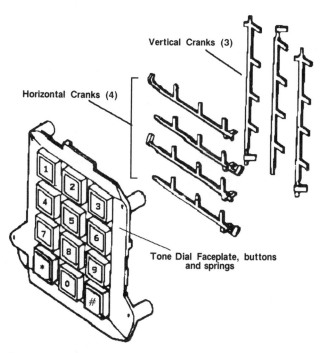

Vertical Cranks (3)

Horizontal Cranks (4)

Tone Dial Faceplate, buttons and springs

Fig. 1-24 *To press one button and have two contacts short together in a matrix type format, telephone designers came up with the horizontal and vertical cranking system.*

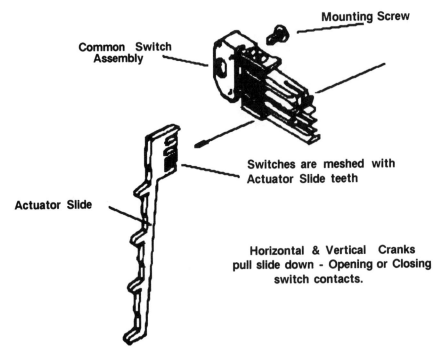

Mounting Screw

Common Switch Assembly

Switches are meshed with Actuator Slide teeth

Actuator Slide

Horizontal & Vertical Cranks pull slide down - Opening or Closing switch contacts.

Fig. 1-25   *To open the common switch assembly*
whenever any button is pressed, designers came up with the actuator slide,
a comblike device that opens and closes the individual contacts on the switch assembly.

36opg). The cost of manufacturing such a circuit, even on the scale of mass production proved to be too high compared to that of the integrated circuit.

ICs that provide a highly stable DTMF output have been introduced in the late 1970s to early 1980s. Like the Plessey's MV5089, ICs of this type produce the required DTMF output with extreme precision. This exactness can be contributed to the use of an internal crystal oscillator that runs at a frequency of 3.58 MHz. With the use of *divider circuits*, this high frequency can be easily transformed into all the single tones needed to create a two-tone telephone audio signal. The signal is just like the conventional I/C (inductor/capacitor) tone dials except for one big difference.

The difference can be seen if you compare Fig. 1-28 with that of Fig. 1-19. Figures 1-19 illustrates a pattern of a pure sine wave. Notice the smooth, even flow as the wave progresses through its normal cycle. Now look at Fig. 1-28. This is an output created by a DTMF generator IC. Notice the *staircase* type of pattern. This is generated by a D/A (digital-to-analog) conversion circuit deep within the IC itself. The D/A converter makes use of a circuit called a R-2R ladder network that produces slight voltage changes with different inputs of binary data. The results is illustrated in Fig. 1-28.

DTMF ICs can produce, digitally, sine waves that are so accurate in frequency and symmetry that conversion circuitry at the central office end of the network cannot tell the difference between the two.

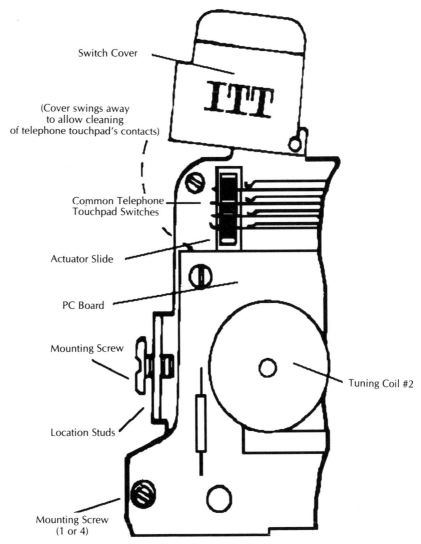

Fig. 1-26 *A partial view of the 32G tone dial.*
Note the location of the actuator slide and the common switch assembly.

A typical example of a digital tone dial is shown in Fig. 1-29; the 42opg. Unlike its predecessor the 32G, the 42opg makes use of the new technology that is available to the telcom industry. Instead of using the awkward mechanical matrix button assembly, the 42 uses a newly developed flexible membrane switch. Figure 1-30 illustrates how this new technology works.

Using a standard tone dial faceplate and twelve buttons, each button is held suspended above a specially etched printed circuit board by a flexible bubble (membrane). At the top of this bubble is a button of conductive material, usually silicon. When the desired button is pressed, the flexible bubble collapses. This action allows the silicon button to come in contact with the PC

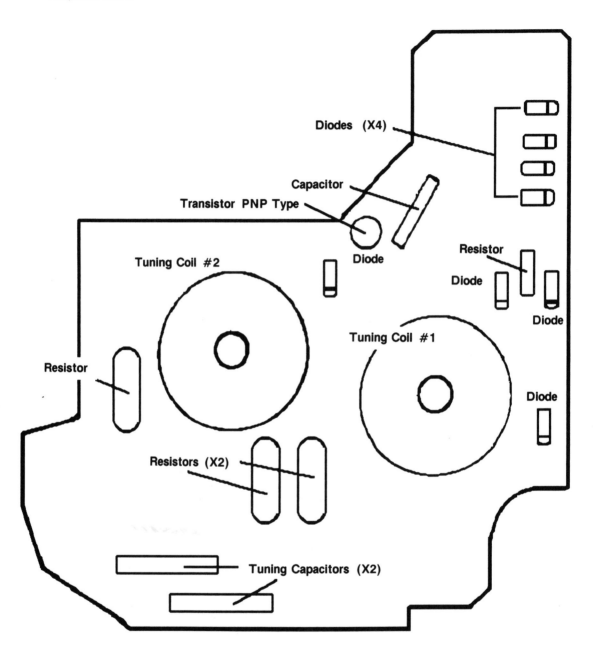

Diodes (X4)

Capacitor

Transistor PNP Type

Diode

Resistor

Diode

Tuning Coil #2

Diode

Tuning Coil #1

Resistor

Diode

Resistors (X2)

Tuning Capacitors (X2)

32G Touch Tone Dial

Fig. 1-27   *The parts layout for a 32G tone dial. Note the use of bulky tuning coils (#1 and #2).*

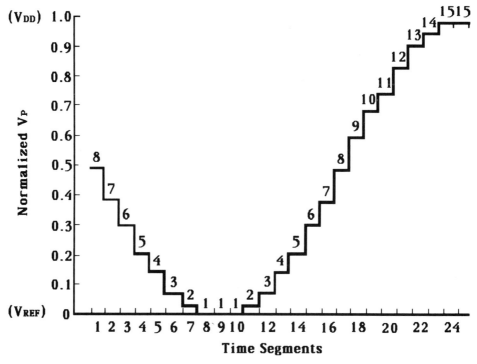

Fig. 1-28 *Technology has advanced to a stage that the bulky tuning coils of the 32G have been replaced by digital tone generation. Above is a standard digital sine wave created by an integrated circuit.*

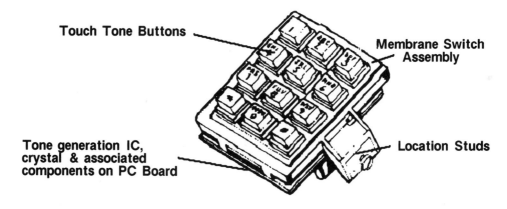

42opg Touch Tone Dial

Fig. 1-29 *To make use of the new digital technology, designers have come up with the 42opg tone dial.*

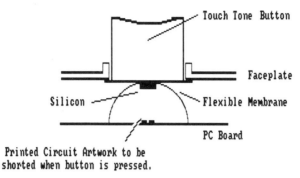

Printed Circuit Artwork to be
shorted when button is pressed.

Fig. 1-30 *Unlike the mechanical cranks used in the older 32 and 36G tone dials, the newer 42opg uses a flexible membrane to make electrical contact.*

board *land* (contact), thus shorting out adjacent *traces*. One trace is connected to the IC row input and the second to the column input. This arrangement, in effect, simulates the twisting of the horizontal and vertical cranks associated with the mechanical tone dial.

Depending on the manufacturer of the DTMF IC, some require that the row and column terminals of the desired output tone be shorted to ground. Some require the connecting together of the row and column terminals. To determine which type you might come in contact with, refer to the manufacturer's specification sheet.

TELEPHONE CORDS

Time has a way of creeping up on people, and everyone looks for an easier way of doing things. Telcom products change because of this desire for improvement. Earlier models came with only one type of line and coiled cords. They were the spade-lug type. These fork-shaped pieces of metal were attached to the ends of wires using a tool called a *crimper*. Then, when installed by the phone company, these lug-tipped wires were screwed to a terminal box located on the wall. If for any reason, the telephone malfunctioned or needed to be replaced, the serviceman had to unscrew the terminal box cover and then unscrew the screws holding the line cord. It really does not seem like a large amount of work, but if you are paying an installer by the hour, this procedure can run into a lot of money. To make the service call as fast as possible, *modular* line and coiled cords were developed. Figure 1-31 illustrates a number of modular as well as the standard spade-lug cords.

To replace a complete telephone or just the handset, all you do is unplug the modular clip from its mating connector and reconnect a replacement device. Easier than before? You bet it is. Modular replacement cords can be found in almost any department store or neighborhood Radio Shack store at a moderate price. But you must be careful to purchase the correct cord type because there are three different types of line cords as well as coiled cords.

Figure 1-31 can also illustrate the different types of cords now available to the consumer. Line cord A (Fig. 1-31) is called *full modular*. Notice the small

Telephone Line Cords

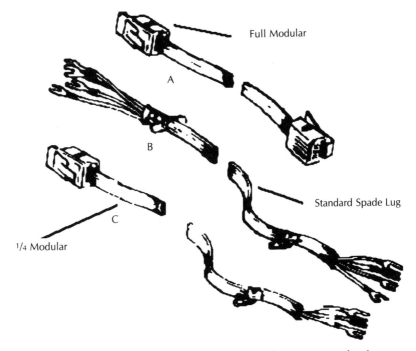

Full Modular

A

B

Standard Spade Lug

C

1/4 Modular

Fig. 1-31 *Telephone line cords come in a wide assortment of styles.*
They range from the older spade-lug cords to the newer full and 1/4 modular types.

modular clips of both ends of the cord. Figure 1-31 also shows the spade-lug type. This cord has lugs on both ends, no clips. Also note the 1/4 modular cord. With this type, spade lugs are on one end, and a modular clip is on the other.

The full modular (Fig. 1-32) and the spade-lug cords also come in a coiled cord model. However, 1/4 modular coil cords are used but not to any great extent. So do not try to find this type in a local store. They just will not be there.

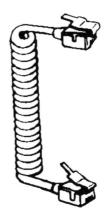

Fig. 1-32 *To allow the replacement of defective telephone handsets easier, service technicians now rely on the standard modular coil cord.*

COMMON TELEPHONE INSTALLATIONS

Before discussing any telephone projects, it is a good idea to stop and briefly go over telephone and project installation techniques.

From the outside telephone connection, installers have a four-wire cable running throughout your house and ended at junction boxes usually at the floor level of rooms where phone service is required. Although the cable contains a red, green, black, and a yellow wire in it, only the red and green are used. The remaining two wires can be connected to the junction box on the wall, but it is really not necessary. They just are not used. Figure 1-33 shows a typical telephone installation from its origin outside the house to its termination inside the telephone.

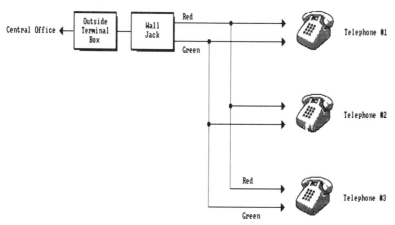

Fig. 1-33 *A block diagram illustrating the connection of more than one telephone to a common telephone line.*

The installation of telephone instruments is a very easy task. Just remember to avoid shorting the red and green wires and observe the line polarity. In other words, connect all red wires to red and green to green. On a rotary line, polarity is no problem. Pulsing of the line will occur whether it takes place on the red or green wire. Tone dials are another story. Remember from the discussion on the requirements of the older 32 and 36G type dials? They require a certain polarity to operate correctly. That is the reason later models have four diodes incorporated in their circuitry (32opg and 36opg). To avoid confusing and the rewiring of an installation, why not observe the line polarity right from the start? Connect red to red and green to green whether the line is for a rotary or tone-dial phone.

Figure 1-34 depicts three types of installations for the projects in this book. View A shows the direct connection, B shows a project in parallel with an existent telephone, and C shows that a project is connected in series with a telephone. Parallel connections are usually used if your design must sense an incoming ring or DTMF signal. Series connections can be used if your project must know if a telephone is taken off the hook. A fourth installation, shown as

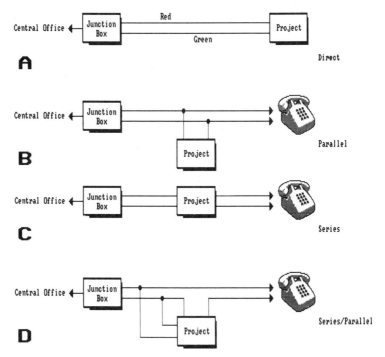

Fig. 1-34 *To connect your telephone projects to the line (A), you might need them to be connected in parallel (B) or in series (C), or even in a combination of both (D).*

Fig. 1-34D, takes into consideration that you might need both a series and parallel connection. An installation of this type can be found with the Tele-Guard II (digital telephone lock) circuit.

An easy way to install a working telephone project to the line is to purchase what is called a *TEE* adapter (Fig. 1-35). This device just snaps into a standard modular wall jack just like a line cord. But with one big difference.

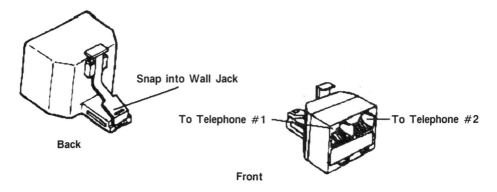

Fig. 1-35 *An easy way to connect two telephones in parallel to a phone line is to use the telephone TEE adapter.*

The TEE adapter allows two telephones or devices to be connected to the same jack. By using the TEE, you can save the time and trouble in wiring up a parallel connection to a telcom project or phone. Adapters of this type can be purchased in any Radio Shack store. For the price of a few dollars, this invention can save you the aggravation of making an additional connection.

When installing a telephone or even a project to a line, just remember two important things:

- DO NOT SHORT THE RED AND GREEN WIRES TOGETHER
  Also you might want to take an ohmmeter and test the project before connecting it to an active telephone line.
- CONNECT ALL RED WIRES TO RED AND ALL GREENS TO GREEN
  Or in other words, observe polarity.

If you follow these simple rules, you should not encounter any difficulties with the installation of any project to the line.

Now that you have a firm grasp on the concepts of basic telecommunications, take the time to consider the type of electronic test equipment and tools that any knowledgeable electronic hobbyist, technician, or even an engineer needs. This equipment is called the tools of the trade and is your next stop.

Tools of the Trade

WHATEVER YOUR OCCUPATION, BE IT A PLUMBER, ELECTRICIAN, STOCK MARKET trader, or electronic engineer (or technician), you must purchase items that help you perform your daily tasks with ease and accuracy. Once the basic tools of the trade are purchased, you should also consider speciality items.

For example, consider a computer enthusiast. The basic tool of the trade is his or her computer system. Then at any time, special enhancement hardware can be purchased and easily added to the system depending upon the field of computer sciences the enthusiast wishes to explore. Desktop publishing is a perfect example of this specialized computer field. Users of this software most likely will purchase, in addition to a basic computer system, an image scanner, laser printer, and maybe additional memory. On the other hand, accountants have no need for a scanner but they can make use of computerized spread sheets and other business accounting software. This scheme can go on and on. Electronic technicians, like computer enthusiasts, are faced with the monumental task of deciding which tools of the trade are needed and purchasing the items at a price that is the most pleasing to their pocketbook. Whether you decide to enter the analog or digital end of electronics, basic tools are required.

This chapter helps you with this chore. The basic electronic tools and test equipment that you might consider adding to your ever-growing work shop is presented and their use is explained.

The novice should consider purchasing inexpensive tools as well as inexpensive test equipment in the beginning. Then as knowledge of electronics increases and expertise develops, you can add more expensive and accurate test equipment and tools to your research and development expense account.

SOLDERING IRONS

Soldering irons (or soldering pencils) come in a wide variety of styles, sizes (wattages), and shapes. These tools are shown in Figs. 2-1 and 2-2. Soldering electronic components to printed circuit boards or copper wires is an art that must be acquired. If you are a beginner, practice soldering before considering actual project construction. This is a must because most electrical components are heat sensitive. Any excess heat generated by soldering can damage or even destroy the delicate internal circuitry of the device. This danger is particularly critical for integrated circuits. Integrated circuits are notorious for being

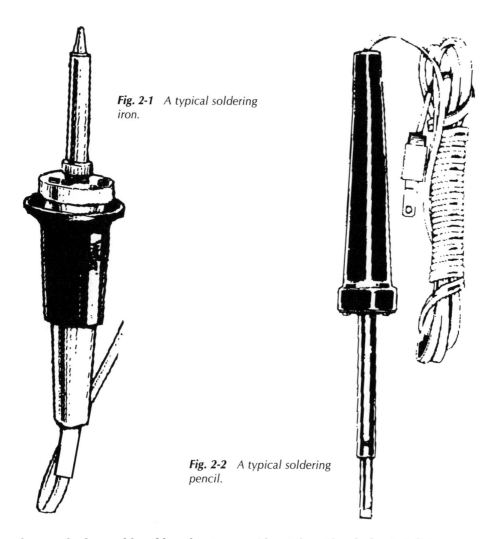

Fig. 2-1 *A typical soldering iron.*

Fig. 2-2 *A typical soldering pencil.*

damaged when soldered by a beginner without the aide of a heat sink (more on that a little later). For this reason, beginners should consider purchasing soldering irons (or pencils) with a rating of about 20 or so watts. These devices generate just the right amount of heat to transform the rosin core solder into liquid. But as stated, if you have limited soldering skills and you wish to build a number of projects presented in this book, consider soldering pieces of bare copper wires together before any assembly. This will allow you to gain that confidence and expertise in soldering. Integrated circuits like the National Semiconductors Digitalker, are relatively expensive items and they are very susceptible to heat damage if soldered incorrectly.

   Good soldering practices are a must when building electronic circuits. It is most frustrating when you spend hours assembling a project then find out (while troubleshooting) that you have unintentionally destroyed one or more components by using careless soldering procedures or by using a soldering iron whose wattage was too high. If this scenario occurs too many times for you, purchase a book on basic electronics procedures and familiarize yourself with soldering techniques.

As stated previously, soldering irons come in a wide variety of styles and wattages. If you plan to do extensive assembly work with electron tubes, a soldering iron of 35 W (watts) or lower is not a wise choice. You might want to consider purchasing a soldering gun. As illustrated in Fig. 2-3, a soldering gun resembles a pistol. It has a handle, a trigger and a specialized tip that heats up when the trigger is pressed. The heat generated by this current-hungry device is tremendous, so using this type of iron on delicate components like ICs and diodes would be foolhardy to say the least. So if you prefer to do extensive assembly using integrated circuits, stick with a low-wattage soldering iron or the soldering pencil.

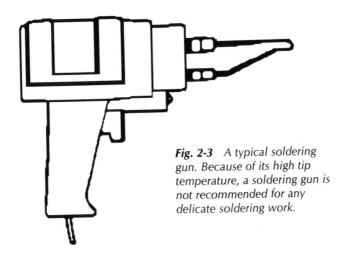

Fig. 2-3 *A typical soldering gun. Because of its high tip temperature, a soldering gun is not recommended for any delicate soldering work.*

SOLDERING IRON STATION

A device that is readily available from electronic mail order houses at a modest cost is the soldering station (see Fig. 2-4). This compact unit not only provides a safe storage place for a hot iron, it also provides a small tin plate that can hold a wet sponge used to clean the soldering tip periodically. But the most important feature of this device is the heat-control knob. This knob governs how hot the tip of the iron will become for any given soldering job. Using a device like this, professional solderer's can accomplish a number of soldering tasks requiring different wattages or heat settings. All this can be achieved using one 100 W soldering iron.

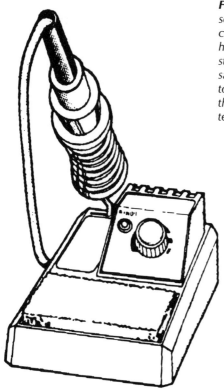

Fig. 2-4 To help the user in soldering electronic components, manufacturers have designed soldering stations. This station provides a safe soldering iron holder, a tray to hold a wet sponge, and a thermostat to control the tip temperature.

A less expensive station can be purchased for about $7.00 (see Fig. 2-5). This device, of course, will not have the temperature control knob as the high-cost unit, but it does provide the tin tray for your wet sponge and a safety holder for the iron. This $7.00 item should be part of your electronic tool chest.

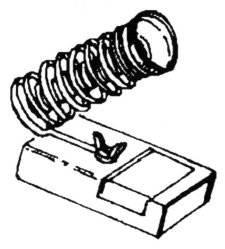

Fig. 2-5 A more inexpensive soldering holder.

A high-cost soldering station with the advanced temperature control is a specialty item usually reserved for professional, industrial solderers. For the home tinkerer, a soldering pencil with a wattage of about 20 W should be purchased for all delicate soldering. More heavy duty requirements (of course this does not include the soldering of ICs, diodes, and other heat-sensitive electronic components) should be fulfilled with a 120 W soldering gun.

SOLDERING IRON TIPS

Once you have decided on the make and model soldering pencil you wish to make use of extensively, soldering tips is the next subject you should consider. As there are a number of styles of soldering irons on the market, so there are an equal number of soldering tips.

Figure 2-6 illustrates three of the most common types of soldering tips available. They are the spade tip (Fig. 2-6A), chisel (Fig. 2-6B) and the needle (Fig. 2-6C). Which of the three types presented you will select depends on the application you have in mind. If you assemble electronic circuits that have large spacing between components or connections, the chisel tip should perform admirably. But if IC construction is your specialty, the needle tip is highly recommended because the pin spacing of an IC is only 0.1 inch. If you use a soldering pencil tip any wider than 0.1 inch, you run the risk of inadvertently creating *solder bridges* between pins. The bridges will create extensive problems because of the difficulty of removing the excessive solder (more on that a little later in this chapter) from the printed circuit board.

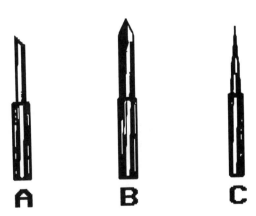

A   **B**   **C**

Fig. 2-6 *The three most commonly used soldering tips. Spade tip (A), chisel (B), and needle-point tip (C).*

To complicate matters a little further, soldering tips also come in a variety of plating configurations. Relatively inexpensive tips are made of copper. When used, copper tips require you to clean oxidation from the tip at regular intervals. They also require frequent replacement. Longer-life tips are available, but they cost a dollar or two more. These more expensive tips are plated with a coating of iron. This iron plating not only gives the tip a longer life, but it also prevents the unsightly oxidation from forming.

Not only can soldering iron tips be found in copper and iron-plated styles, they also come with a coating of nickel (nickel plated). Somewhat cheaper than the iron tip, nickel plating can provide better joints, not only are the joints more appealing to the eye, but they are electrically sound connections as well.

SOLDER

Solder is a component of electronics that people do not give much thought to. If the wrong solder were purchased, disastrous results can be guaranteed. Never use acid-core solder on any electronic project or assembly. This type of solder is a corrosive. Just as demonstrated in a science class you might have attended in school, acid will eat through metal. The same holds true if you use acid-core solder on an electronic circuit. You might not see any problem at first, but give it a little time. Maybe after a few weeks or even a few days, the acid contained in the solder will begin to eat its way through the metal, rendering the project useless. If this were the case, you might as well throw the circuit away. Do not even bother troubleshooting the project. It is a lost cause.

To prevent this from happening to you, rosin-core solder should be the only solder used with electrical and electronic work. Kester, a name associated with high quality solder and chemicals for a number of years (see Fig. 2-7), makes available, not only to industry but also to hobbyists, a wide range of high-quality rosin-core solders. The most commonly used solder in electronics is 60/40 (60 percent tin and 40 percent lead alloy) or Kester 44 rosin core. The Kester 44 Rosin Core Solder is specifically designed for all electrical and electronic applications. The flux residue is completely noncorrosive compared to the highly destructive acid-core solder. Kester 44 makes soldering faster than ever before and provides a stronger, longer-lasting bond. Solder of this type is available on 1-, 4-, and even 5-pound spools in diameters of $3/32$-inch (0.093), $1/16$-inch (0.062), and other popular sizes.

Fig. 2-7 High-quality rosin-cord solder is the only material that should be used for any electronic project.

Kester also makes available computer/electronic solder. This solder, with its cored flux, was designed for electronic soldering applications. The unique flux action combined with a low melting point prevents unwanted heat buildup on parts. This prevents component damage or destruction.

The pricing of solder, whether it is Kester or any other brand, can range from $1.00 for a small length, to a high of $50.00 for a 4-pound spool. For the average electronic tinkerer, a 1-pound spool of 60/40 rosin-core solder can last for quite a long time.

DESOLDERING TOOLS

Earlier, I mentioned that using a soldering pencil with too wide a point can cause unwanted solder bridges between electrical conductors. Removing these bridges can be a monumental task if you do not have the proper tools.

Without the customary implements for removal, the beginner might be tempted to eliminate bridges by first melting the solder that is causing the problem then *banging* the PC board on its side to force the solder to splash from the component onto the table.

This is the most convenient method, but can create other problems by splashing molten solder onto other perfectly soldered connections, resulting in another two or three solder bridges.

Not to create unnecessary troubleshooting time for yourself, consider three items designed specifically to remove unwanted solder safely from circuit boards or chassis.

Figure 2-8 illustrates a *desoldering bulb*. The principle of operation is quite simple. Just melt the unwanted solder using your iron, then squeeze the bulb. Placing the bulb tip close to the molten solder, release the bulb and allow it to expand into its normal shape. It is this expansion that creates a vacuum at the tip. This vacuum is the force that literally sucks up the molten solder from the connection to eliminate the solder bridge. A device of this type can be purchased from a local Radio Shack store or electronic mail order house for about $2.00 to $3.00.

Fig. 2-8 *An inexpensive way to remove solder from a connection is to use the desoldering bulb.*

Another method of desoldering is shown in Fig. 2-9. A *desoldering pump* (or solder sucker as it is called by technicians), also produces a vacuum but in a different way. Using the principle of a spring-loaded piston, a highly affective vacuum can be made. The pison is located within the hollow handle of the solder sucker. To produce an air-tight seal, a rubber O-ring is installed around the piston. The piston also has a metal shaft connected to its center. Around this shaft is a spring. It is this spring that is compressed, causing the piston to be pushed near the tip of the desoldering pump, where it is locked into place. When released by the thumb, the spring pops up into its normal resting position. It is this action that forces the piston upward to cause a powerful surge or air to enter through the opening of the tip. It is this surge of air that produces the needed vacuum that removes the molten solder from a connection point.

Desoldering pumps come in a broad range of styles and prices. Inexpensive pumps are constructed from plastic, and they do not last very long. The more expensive models are made from metal. Whichever type of pump you decide to purchase, desoldering tips are frequently replaced. They are only made of nylon. If touched by the hot point of the soldering iron or pencil, they will melt. But the price of replacement tips is extremely low. So buy two or three to keep on hand.

Fig. 2-9 *The desoldering pump (or solder sucker) removes solder and is more expensive than the desoldering bulb.*

Technology has incorporated two completely separate electronic items into one tool, the soldering iron and desoldering pump. Figure 2-10 depicts what manufacturers have developed. This combination solderer and desolderer provides the best of two worlds. This device combines the popular snap-action vacuum solder removal tool with a soldering iron for fast, one-handed operation. A device of this type is ideal for field repairs and even production work. The desoldering iron has a tightly sealed vacuum pump that snaps up all molten solder quickly and easily. The device also has a removable

Fig. 2-10 *The combination soldering iron and desoldering pump is more expensive than the desoldering pump.*

reservoir that stores the discarded solder. This reservoir must be cleaned periodically to prevent clogging. The price for this one-handed convenience is about $25.00. A tool of this type can be classified as one of those specialty items. So if you are a beginner in electronics, do not run out to buy one. At this time, it really is not necessary.

SOLDERING HEAT SINKS

Time and time again, the rule "While soldering heat sensitive components, use a *heat sink*" is repeated. So put this recommendation to rest once and for all. Depicted in Fig. 2-11 are two devices used to draw away the damaging heat caused by soldering irons. Figure 2-11A illustrates a standard "Spring Loaded" heat sink, and Fig. 2-11B shows a surgical clamp (or *hemostat*). Both

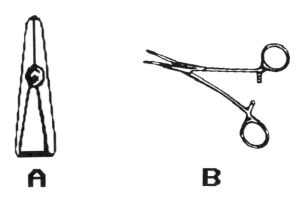

A   **B**

Fig. 2-11 *To draw heat away from delicate electronic components, use heat sinks such as the common spring-loaded heat sink (A) depicts a surgical clamp (or hemostat) (B).*

items are clipped between the point to be soldered and the component. This clipping action draws the heat of soldering from the component. When soldering is complete, you must realize that the heat sinks will be hot to the touch. So use caution when removing the heat sink.

Also a standard alligator clip used on test leads can also be used as a heat sink, just make sure that the clip is small enough to properly grab the lead or IC pin so it can protect the device.

WIRE STRIPPERS, PLIERS, AND CUTTERS

A chapter on electronic tools of the trade can not be complete without mentioning the three most basic accessories to any would-be engineer's tool box. And that is the wire strippers (see Fig. 2-12), long-nose pliers and the diagonal cutters (see Fig. 2-13).

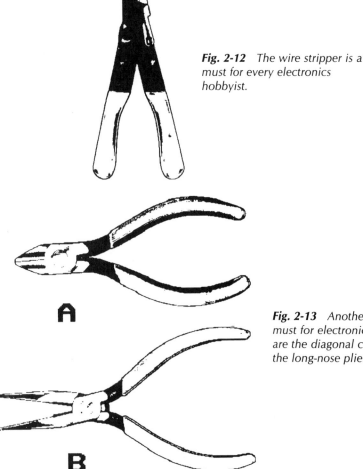

Fig. 2-12 *The wire stripper is a must for every electronics hobbyist.*

Fig. 2-13 *Another absolute must for electronic hobbyists are the diagonal cutters (A) and the long-nose pliers (B).*

Wire strippers, as the name applies, strips or removes the protective insulation that surrounds copper wires and cables. As with everything else in electronics, wire strippers come in a variety of styles. These styles can handle any type of job you can think of. But for your basic tool box, concentrate your efforts on strippers that can easily remove the insulation of electrical hookup wire. Selling for a price of about $4.00 to $6.00, strippers come with an adjustable screw stop. This stop is used to prevent the cutting edge of the strippers from nicking the copper wire. The mechanism is usually calibrated for the thickness of the wire (or its *gauge*). Number 22 AWG (American Wire Gauge) is the most commonly used wire in electronic project assembly. Wire strippers are a relatively inexpensive addition to any tool box, and a stripper should be in yours.

Figure 2-13 illustrates the other common electronic tools. The diagonal cutters (Fig. 2-13A) and the long-nose pliers (Fig. 2-13B). Diagonal cutters are used to cut excess component leads upon the completion of soldering the part to a printed circuit board. Variations in cutter style, length, and cutting angle are commonplace and should be taken into consideration when purchasing the item.

Long-nose pliers are used to hold and manipulate small components and wires while stuffing them on a PC board or while soldering. As with the diagonal cutters, long-nose pliers also come in a wide variety of styles and nose lengths. Only trial and error will allow you to determine which type of pliers and cutters best suits your needs. For this reason, beginners should purchase very inexpensive pliers and cutters, as well as wire strippers until you know what requirements are needed from the tools.

THE ALTERNATIVE: THE ELECTRONIC TOOL KIT

A number of tools that every engineer or technician should have in their tool box have been discussed. Obtaining these items individually may prove to be too expensive, especially if you buy them from mail order houses that require a minimum purchase.

As an alternative to this large expenditure, you might consider purchasing an electronic tool kit (see Fig. 2-14). These kits contain all the basic tools required by any technician. With a modest layout of cash, you can obtain a soldering pencil, wire strippers, diagonal cutters, long-nose pliers, solder, and a set of screwdrivers. For those just starting out in the exciting field of electronics, an electronic tool kit from your favorite mail order house is highly recommended.

PRINTED CIRCUIT BOARD HOLDER

Another one of those specialty items is shown in Fig. 2-15. This device holds a printed circuit board between its two protruding arms so that soldering and component mounting can be easily accomplished. Again, a large number of variations in design are available from just about any store that caters to the electronic enthusiast.

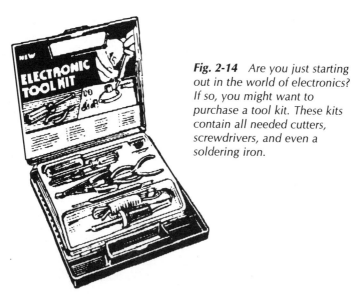

Fig. 2-14 *Are you just starting out in the world of electronics? If so, you might want to purchase a tool kit. These kits contain all needed cutters, screwdrivers, and even a soldering iron.*

The model shown in Fig. 2-15 features an eight-position rotating head and six lock positions in the vertical plane. Cross bars are available up to 30 inches long to hold larger printed circuit boards. To expedite PC Board insertion and removal, the holder contains a spring-loaded head and quick board rotation for easy component insertion and soldering. A device of this type usually sells for about $46.00. Less expensive models are also available, but if you have no problem soldering a printed circuit board that is lying upside down on your work table while you solder components, the investment of $46.00 is not mandatory.

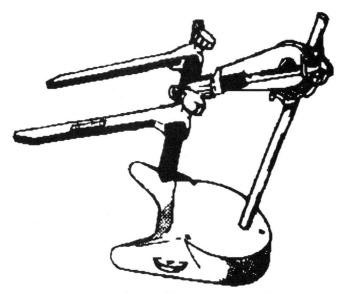

Fig. 2-15 *A luxury for the old timer: a PC board holder.*

DRILLING PRINTED CIRCUIT BOARDS

If you plan to make your own printed circuit boards for your designs, one tool is a must to have: it is the rotary drill (see Fig. 2-16). This hand-held device accepts drill bits too small to be used with the standard 1/4-inch chuck of an electric drill. So by using a hand-held rotary drill, the boring of a 0.042-inch hole in your etched PCB (printed circuit board) for component mounting is extremely fast and easy. A drill of this type can handle other important jobs for you. Just by purchasing the proper rotary bit, you can buff, polish, cut, and sand as well as drill holes. For a price of about $45.00, this instrument will pay for itself in no time at all.

Fig. 2-16 *If you intend to make your own printed circuit boards, you might want to purchase a small electric drill.*

ROTARY HAND-HELD DRILL BITS

Upon purchasing the Rotary Drill, you might want to consider buying an assortment of drill bits. Figure 2-17 illustrates a package of 20 common drill bits used to bore holes in the printed circuit material. The sizes available with

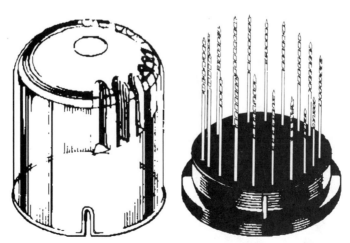

Fig. 2-17 *A supply of drill bits (extra small size—bits of approximately 0.042-inch diameter) are needed.*

this package range from #61 to #81 (#61 being the largest and #81 being the smallest).

When purchasing bits to drill printed circuit boards, you must consider spending at least $2.00 for each. The high cost is due to the carbide-tipped material used in the manufacturing process. Carbide-tipped drills are highly recommended because they contain a very sharp tip that can easily cut into the etched copper without lifting the Land from the circuit board. You can use less expensive drill bits, but you do run the risk of damaging or even destroying the delicate copper trace of the board. To prevent this, you should invest in the more expensive carbide-tipped drill bit.

THE VOM (VOLT-OHM-MILLIAMMETER)

The VOM will, without a doubt, become the most used electronic testing device you will have in your arsenal of troubleshooting equipment. The VOM will not only determine values of resistance, but will also give direct readings of current and voltage. By using a scale selector switch (see Fig. 2-18), you can measure voltages from the most minute units ranging into the thousands of volts (using special *probes*).

The resistance measurements of VOMs are less likely to be as accurate as voltage measurements. With a typical, moderately priced VOM, resistances can be measured within a range of about 10 to 10 MΩ with some degree of efficiency. Of course, higher-priced instruments can provide accuracy that might not be needed by the kitchen-table hobbyist. So before purchasing a VOM, contemplate your future needs. If you feel that the additional money invested in a high-accuracy VOM will pay for itself in the long run, by all means, invest in your future.

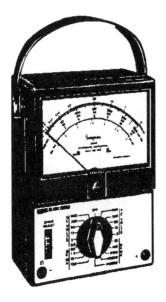

Fig. 2-18 *If you plan to troubleshoot defective circuits, you will need a VOM (volt-ohm-milliammeter).*

THE DVM (DIGITAL VOLTMETER)

Since the advent of LSI (large-scale integration—over 100 components per chip) digital test equipment has been popping up all over the place. Relatively inexpensive resistance, voltage, and current measurement instruments now provide the accuracy that the analog VOM cannot compete with. Figures 2-19 and 2-20 illustrate two types of digital voltmeters available to the hobbyist of modest means. Operating usually from a transistor battery, the DVM makes use of an LCD (liquid-crystal display). By using LCD's, battery life is extended tremendously as compared to the now almost extinct seven-segment light emitting diode display.

Whether you purchase a VOM or DVM for your troubleshooting work, the choice is really up to you. Buying a DVM for voltage measurements would be a good deal. But for checking the charging and discharging of capacitors or the leakage of a diode, the old analog predecessor is better.

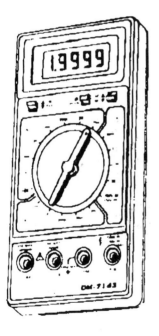

Fig. 2-19   DVM (digital voltmeters) are great to have around but they are expensive. They might be too expensive for the hobbyist.

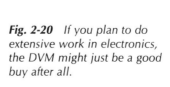

Fig. 2-20   If you plan to do extensive work in electronics, the DVM might just be a good buy after all.

OSCILLOSCOPES

Like VOMs, oscilloscopes must be purchased with a number of factors in mind. You should ask yourself if most of your troubleshooting will be restricted to the digital or audio spectrums. If audio is your expertise, consider purchasing an oscilloscope with a frequency response from 10 to 20,000 Hz. These scopes make use of what is called *ac coupling* and are relatively inexpensive. Although a digital scope can make use of a frequency response from a pure dc voltage to millions of cycles or pulses per second, it goes without saying that the higher the frequency response of a particular digital oscilloscope, the price tag rises proportionally.

Figure 2-21 illustrates a typical digital oscilloscope which makes use of this extremely high frequency response. It also contains dual-trace and triggering capability. Through the advancements of electronics in the world today, scopes can now be placed in a category with computers. Microprocessor-controlled oscilloscopes are a common sight in many research and development departments of large manufacturing and research corporations.

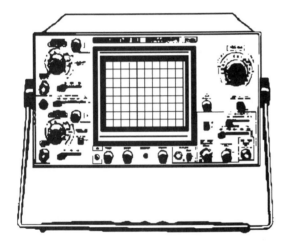

Fig. 2-21 *A must for any electronic hobbyist is the oscilloscope. Be aware of your troubleshooting requirements; then purchase the appropriate oscilloscope that fits those needs.*

The scopes available to the hobbyist can make one throw up their hands in frustration. But you are not defeated. Not just yet. Consider building your own. Heath, a company with years of experience dealing with the electronic hobbyist, offers very high quality oscilloscopes in kit form at reasonable prices. These Heathkits will not only give you an attractive, accurate scope, but you also will gain the invaluable experience of assembling a complicated piece of gear. Do not let the word complicated prevent you from purchasing the scope. Heathkit has the reputation of providing easy step-by-step illustrated instructions with every kit. Heath also provides a troubleshooting service for those of you who just cannot find that solder splash in the vertical oscillator. So you are guaranteed a priceless piece of test gear when construction is finally complete.

INSERTION TOOL

Figure 2-22 depicts an insertion tool, which is a must if you plan to do much telephone work. This device is a tool that securely holds a wired spade lug at its tip. This tool will allow the lug to be easily pushed into the correct terminal of a tightly packed printed circuit board. Termination clips of the type just mentioned are widely used inside your telephone (just take a look at the network inside the phone). If you plan to construct many of the telephone projects illustrated in this book, you will find that spade lugs can deliver painful finger cuts if not inserted correctly. So to prevent unscheduled trips to the medicine cabinet for a bandage, make the spade lug insertion tool part of your tool box.

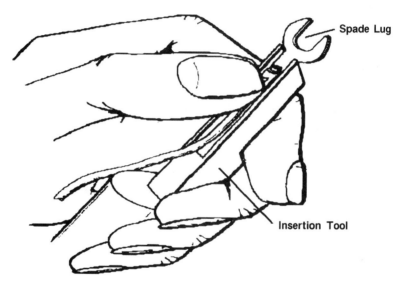

Fig. 2-22 *Doing a lot of telephone design work? You might want to purchase a spade-lug insertion tool.*

You might ask, Where can I buy one? Well tools of this type cannot be found in your local Radio Shack store, but if you subscribe to electronic magazines like *Radio-Electronics*, *Modern Electronics*, or *Popular Electronics*, you might have noticed advertisements from companies that deals exclusively with hard-to-find tools. Write to these businesses and request a current catalog or flyer. You might be in luck and find exactly what the doctor ordered.

3

Teltone Corporation Telcom Chips

M ANY OF THE PROJECTS IN THIS BOOK USE INTEGRATED CIRCUITS THAT SENSE AND decode the DTMF and rotary pulses. To discuss the operation and functions of the available ICs in detail would require a book in itself. Rather than presenting page after page of boring theory, a condensed pamphlet is reprinted. A major manufacturer of telcom chips, Teltone Corporation of Kirkland, Washington, was gracious enough to allow reprint of their SC-1 pamphlet.

This chapter provides the needed information and pinout diagrams on a number of chips available from Teltone. This information can be put to use in many circuits of your own design. In this condensed version, SC-1 illustrates the internal workings of the M-927 pulse counting IC as well as the M-957 DTMF receiver. Computer interfacing techniques also are discussed. Thanks to the Teltone Corporation for their help and for allowing inclusion of their SC-1 pamphlet in this book.

APPLICATIONS FOR DTMF AND PULSE TELEPHONE DIALING INCLUDING REMOTE ACCESS AND CONTROL

---

* Reprinted with permission of Teltone Corporation, Kirkland, Washington.

Computer data entry, operations monitoring, and equipment control are just a few of the many functions that can be performed from remote locations using a standard telephone set and the public switched telephone network. Telephone dialing is a simple yet reliable form of data transmission that offers system designers the following unique advantages:

- A worldwide network already in place and economical to use
- Inexpensive and user-friendly terminals (telephone sets)
- Readily obtainable hardware

Teltone Corporation offers an entire family of DTMF and pulse-dialing receivers to make your application simple and cost effective. Because of these two considerations, extensive use is made of the M-927 (telephone pulse receiver) and the M-957 (DTMF receiver). Both of these instruments are manufactured by Teltone Corporation and can be purchased from Del-Phone Industries, Spring Hills, Florida (see appendix D for a complete list of names and addresses of sources.) Drop Del-Phone a line and request their current catalog listing of available ICs and prices.

This guide provides the background information you will need to get started and the detailed diagrams and descriptions of circuits that you can use, either as is or modified to suit your purposes.

PSTN (PUBLIC SWITCHED TELEPHONE NETWORK)

The principle elements of the public switched telephone network are shown in block diagram form in Fig. 3-1. The telephone line is a wire loop that connects each station to a local switching center called a CO (central office). The called and calling stations can be connected to the same CO, or they can be connected to different COs linked by intermediate switching centers. Two types of station instruments are in common use today: pulse-dialing (rotary) telephones and tone-dialing telephones. Pulse-dialing telephones transmit digits as serial *makes* (connections) and *breaks* (disconnections) of the telephone line. Tone-dialing telephones generate pairs of tones called dual tone multifrequency tones. DTMF dialing is the newer system and is usually preferred, for reasons that will be explained later in this chapter.

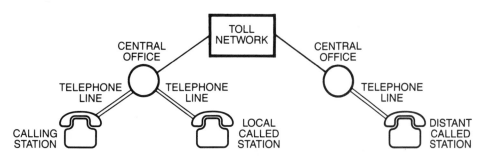

Fig. 3-1   *Public switch telephone network.*

THE TELEPHONE LINE

In Fig. 3-2A, the station-to-CO connection is shown in detail. Lifting the telephone handset is called going *off-hook*, and its effect is to close the telephone hook-switch and allow dc (direct current called *loop current*) to energize the CO current-sense relay. The CO then couples the dial tone to the line, which is the caller's signal to dial. The pulses or tones generated by the calling telephone are detected at the CO and used to route the call. Replacing the telephone handset is going *on-hook*, and its effect is to open the hook-switch, interrupt the flow of loop current, and de-energize the current-sense relay.

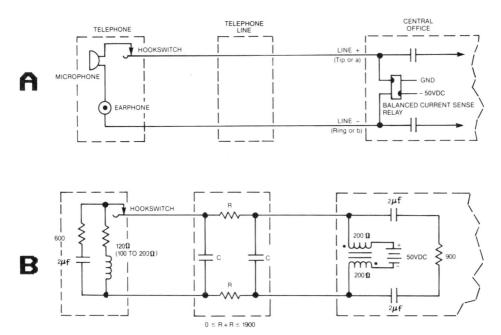

Fig. 3-2   *Telephone line terminations (A) and telephone-line schematic diagram (B).*

After the relay has been de-energized for several hundred milliseconds, the CO drops the connection and the system returns to its idle state.

Figure 3-2B is a schematic of the electrical circuit formed by the elements of Figure 3-2A. In North America, the off-hook telephone appears as about 600 Ω ac and 100 to 200 Ω dc. The telephone line has wire-to-wire capacitance and simple resistance resulting from the length of the wire. Long telephone lines have capacitances that can attenuate the higher frequencies and resistances that can limit the dc current. The most common CO circuit equivalent is a dc battery current through two large, balanced inductances in parallel with a capacitively coupled 900 Ω resistance. The inductances are usually part of the current sense relay, and the battery voltages can be – 36 to – 60 Vdc depending on the system. In North America, – 50 Vdc is the most common voltage.

Figure 3-3 shows the general line conditions that occur during an off-hook and dialing sequence as seen from the CO. DTMF and pulse dialing are included as part of the same sequence, although this is not likely to occur in a real application. Loop current is usually 20 to 100 mA (milliamperes), and DTMF tones arriving at the CO are usually 50 to 1000 mV (rms) depending on line conditions. Most pulse dialing occurs at 8 to 12 PPS (pulses per second) with a line make-to-break ratio of about 60 to 40. This means that a large digit such as 9 can have a duration greater than one second. In contrast, DTMF tones are usually 50 milliseconds in duration with intervals between tones of at least 40 milliseconds, for a total signal cycle time of about 100 milliseconds. The significantly greater speed of DTMF dialing—as much as a factor of 10—is one of its chief advantages over pulse dialing.

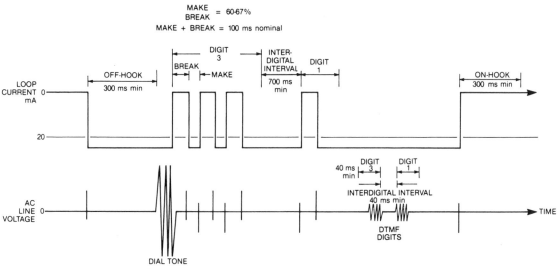

Fig. 3-3   *Telephone line conditions (DTMF and pulse).*

THE STATION-TO-STATION CONNECTION

Figure 3-4A shows a station-to-station connection in which the calling station and the called station are linked via toll transmission facilities. Dialed digits are stored at the CO of the calling station and then forwarded via other switching centers to the CO of the called station. This signalling is either along the switching path as multifrequency (a special 2-of-6 tone code) or via a separate digital channel dedicated to this purpose. When the call is answered, an indication is returned through the network from the called station CO to toll-recording equipment at the calling station CO.

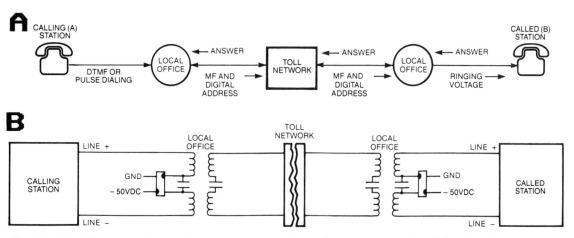

Fig. 3-4   *Signalling on a call to a distant station (A) and typical circuit for complete link with a distant station.*

In Fig. 3-4B the basic station-to-station connection is shown in an electrical schematic. Note the capacitor and transformer coupling, which limits dc signals to the local telephone lines and ensures that only ac signals are transmitted station to station. Originally engineered to provide a channel for speech, this ac path also accommodates newer schemes for communicating in the same frequency spectrum, extending from 300 to 3300 Hz. These schemes include facsimile and DTMF. The capability for station-to-station signalling, also called end-to-end signalling, is another advantage of DTMF dialing over pulse dialing.

Although the human ear can tolerate quite a bit of noise and still understand spoken words, the electronic detection of spectral information is complicated by impulse noise, random noise, cross-channel interference, and circuit losses. The biggest penalties are imposed by the local telephone lines where losses due to line resistance and capacitance are often as great as 9 dB (decibels). Toll transmission facilities have better control over conditions and usually do not impose more than a 10-decibel loss, despite their longer length. Both toll and local lines also introduce noise of many types. As shown in Fig. 3-5, a tone transmitted at one station might be greatly attenuated at the receiving station. Moreover, there is likely to be additional difficulty for the receiver because of signal-to-noise ratio and individual tone-level variations. For these reasons, the engineering of DTMF receivers usually involves compromises and design trade offs between sensitivity and detection reliability.

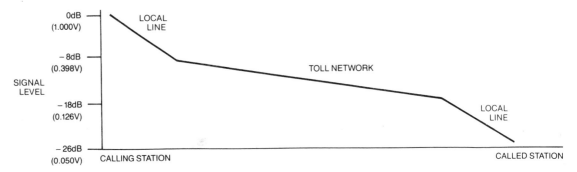

Fig. 3-5 *Example of losses encountered over a link.*

DTMF (DUAL TONE MULTIFREQUENCY) SIGNALLING

The DTMF dialing scheme was developed by Bell Laboratories and introduced in the United States in the mid 1960s as an alternative to pulse (rotary) dialing. Offering increased speed, improved reliability, and the convenience of end-to-end signaling, DTMF has since been adopted as a standard and recommended for use by the CCITT (International Telephone and Telegraph Consultative Committee), CEPT (Conference of European Postal Telecommunications Administrations), NTTPC (Nippon Telegraph and Telephone), and others around the world.

Each of the possible DTMF signals shown in Fig. 3-6 is a composite of one frequency from a high-frequency (column) group and one frequency from the low-frequency (row) group. This 2-of-8 scheme is often reduced to 2-of-7 because signals in the 1633 Hz column are reserved for special nondialing functions.

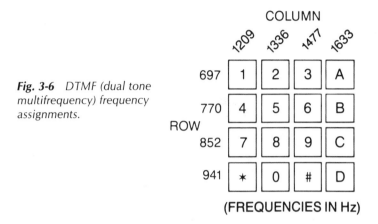

Fig. 3-6   *DTMF (dual tone multifrequency) frequency assignments.*

The divisions of frequencies into high and low groups simplifies the design of a DTMF receiver as shown in Fig. 3-7. This particular design (Teltone from approximately the late 1970s) uses a standard approach. When connected to a telephone line, radio receiver, or other DTMF signal source, the receiver filters out dial tone and noise, separates the signal into its high-frequency group and low-frequency group components, and then digitally measures zero crossings over averaging periods to produce digit decoding.

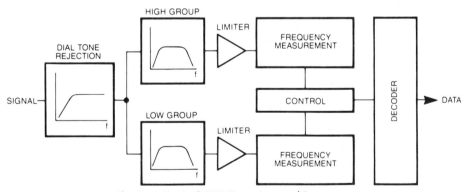

Fig. 3-7   *Typical DTMF receiver architecture.*

As shown in Fig. 3-8, the detection of DTMF signals can be complicated by the presence of 50/60 Hz power line noise, dial tones of various frequencies, random noise, and other sources of interference. Dealing with these problems while remaining immune to speech-simulated digits presents the greatest challenge to DTMF engineers. The tolerable interference line shown in the figure is

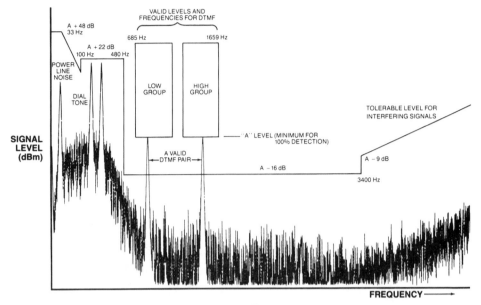

Fig. 3-8   *DTMF signalling environment.*

recommended by the CEPT and is considered a design goal by Teltone and other manufacturers of quality DTMF receivers.

DETECTING SIGNALS AS THE PRIMARY RECEIVER

DTMF and pulse dialing receivers were originally developed as *primary* or dialed digit receivers whose data was used to determine call destinations. Examples of their use in this capacity, is presented in this chapter. Figure 3-9 shows a typical PBX (private branch exchange) or CO switching equipment application, and Fig. 3-10 shows a simple receiver connected to a telephone set for dedicated digit entry. Both circuits are easy to apply because of the wide dynamic ranges and time-guarded input at pin 18 of the M-917 IC. For more detailed specifications, you can write to Teltone Corporation and request data sheets for each receiver integrated circuit. They would be more than happy to supply this information to you.

The critical parameters in the circuit of Fig. 3-9 are good line balance, low insertion loss, and light line loading. These are easily achieved with the Teltone M-927 receiver and the M-949 balanced line-sense relay. The M-949 is energized by the flow of loop current that results when a caller goes off-hook. This current drives the input pin 13 of the M-927 true and enables the receiver to detect dialed digits. Tying the STROBE (pin 34) output to the DTI (pin 7) and RDI (pin 39) as shown causes the receiver to recognize only DTMF digits or only pulse digits, depending on the type of digit first received. When loop current is lost and the M-949 is de-energized for a prescribed duration, the receiver recognizes on-hook and returns to idle. The signal input of the M-927 is ac

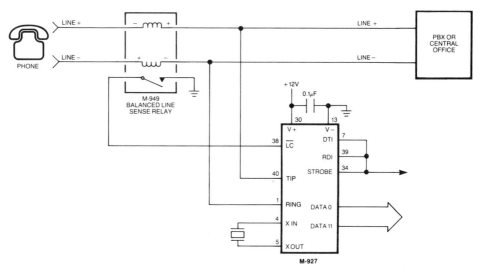

Fig. 3-9   *Tone and pulse line monitor with mode locking.*

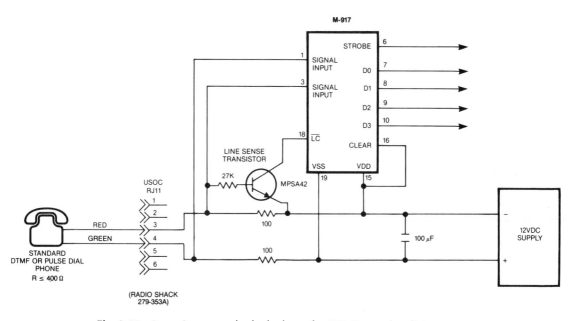

Fig. 3-10   *Powering a standard telephone for DTMF or pulse digit entry.*

coupled and will tolerate ac, dc, and impulse voltages. The M-949 has a minimum of 63 decibels of balance and provides 1500 V isolation from line to relay contact. In the United States, the use of these parts allows registration under FCC (Federal Communications Commission) regulations.

If the application is less demanding, the circuit of Fig. 3-10 might be suitable. Here a standard telephone set is powered by a 12 Vdc supply isolated by two resistors. Across one resistor is a transistor that turns on when the telephone is taken off-hook and turns off when the telephone is put on-hook. The transistor drives the Teltone M-917 input (pin 18), which operates identically to the M-927 relay driven input (pin 38) in Fig. 3-9.

DETECTING SIGNALS AT THE CALLED STATION

It is often useful to be able to dial up a remote station, have the call be answered automatically, and then transmit data using either the DTMF keypad on the telephone or an acoustic DTMF generator. Figure 3-11 shows a typical circuit for detecting ringing voltage and answering the line, and Fig. 3-12 shows a typical circuit for retransmitting received data.

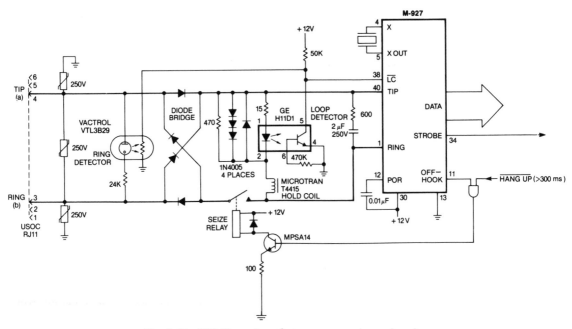

Fig. 3-11 *DTMF receiver that answers an incoming ring.*

In Fig. 3-11, the application of ringing voltage to the telephone line causes the neon isolator to flash (fire), driving the M-927 input true. This in turn drives the M-927 OFF-HOOK (pin 11) output false, which energizes the seize relay and permits dc current to flow, thereby answering the call. Digits dialed by the caller can then be received and decoded. The flow of dc current is maintained by the hold coil, and ac impedance is maintained by a 600 Ω resistor, in series with a capacitor, across the line. Note that the polarity of the CO voltage is important, so a diode bridge should be used if polarity reversals can occur. While the seize relay is energized, an LED opto-isolator monitors the flow of loop current. If loop current is lost for more than 300 milliseconds, because of either the caller's going on-hook or a command on the line shown as HANG UP

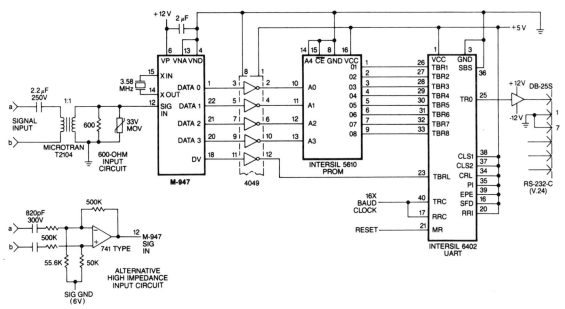

Fig. 3-12 *DTMF-to-asynchronous serial ASCII translation.*

(input to NAND gate connected to the base of the MPSA14 transistor), the system returns to idle until ringing voltage is applied again. The telephone line connector shown is the United States standard RJ11, but could be any type available in your junk box. Protection from common mode or differential line transients is provided by MOVs (metal oxide varistors).

Receivers such as the Teltone M-947 are constructed to use *single-ended signals*, which appear with reference to ground only. Such receivers are easily connected to balanced lines using either of the input circuits shown in Fig. 3-12. The M-949 output is shown converted to a 7-bit ASCII format for asynchronous transmission on an EIA RS-232-C (CCITT V.24) line. Specific types of PROMs (programmable read-only memories) and UARTs (universal asynchronous receiver-transmitters) are indicated, but devices from other manufacturers could also be used.

INTERFACING TO OTHER CIRCUITS

One of the challenges presented to digital designers is inexpensive interfacing to ICs outside the popular standard logic families. Problems are presented in the form of power supplies, inconvenient pinouts, or complex control functions. Fortunately for designers, the Teltone family of DTMF receivers were configured with flexibility and ease of use in mind. Here are some shortcuts and interfacing techniques that can simplify your design and help make it more cost effective.

Figure 3-13 presents various ideas for interfacing a 12 V M-927 receiver with a popular 5 V, single-chip microcomputer, the 8035. The 3.58 MHz clock

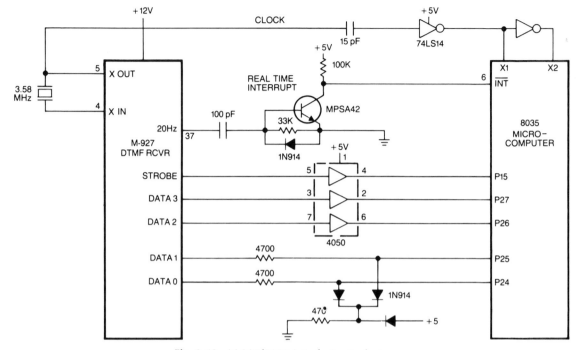

Fig. 3-13 *M-927/8035 interfacing techniques.*

oscillator in the receiver can be used to drive the processor clock, once again eliminating a second crystal. A *real-time* clock is also possible using one of the three M-927 three clock outputs to provide a narrow INT pulse at each rising edge of the square wave. CMOS (complementary metal-oxide semiconductor) 4049 or 4050 ICs are a particularly inexpensive and an easy way to translate 12 V logic to 5 V logic, but diodes and resistors will also work.

Figure 3-14 shows how one crystal-connected M-927 or M-947 can be used to drive the time bases of additional receivers. Use of this technique reduces the parts count and increases system density in multireceiver applications.

Figure 3-15 shows how a simple hardware DTMF digit repeater can be made using the 2-of-8 mode output of the M-927. Once again, CMOS 4049 and

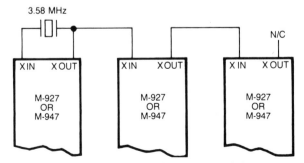

Fig. 3-14 *Cascading clocks for the M-927 and the M-947 IC.*

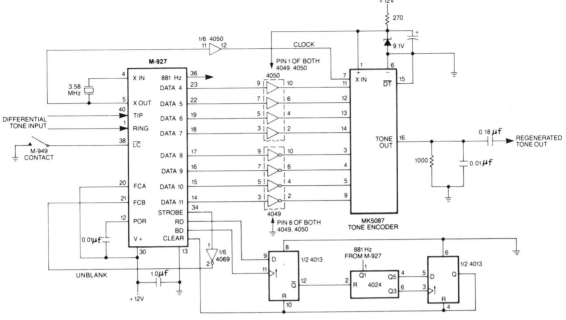

Fig. 3-15   *Pulse-to-DTMF converter and DTMF repeater.*

4050 ICs provide cost effective voltage translation. The M-927 ''pushes the buttons'' on the Mostek MK-5087, regenerating the DTMF tones without the noise and frequency deviation they arrived with. The lower supply voltage of the 5087 is provided by a zener-diode regulator. Other keyboard-compatible products like pulse dialers could be used similarly to provide DTMF-to-pulse translation very inexpensively.

SPECIAL APPLICATIONS

An interesting features of sampled-data processors is that their characteristics change when their clock rates change. Applications of this phenomenon can be seen in Figs. 3-16 and 3-17.

Figure 3-16, a Mostek 5087 DTMF generator and a Teltone M-947 DTMF receiver are operated with 4.43 MHz European color-burst crystals instead of the specified 3.58 MHz North American color-burst crystals. The higher clock rate causes the generator output tones to be higher in frequency and the receiver recognition frequencies to be shifted by about the same proportion. The overall effect is to provide tone signalling at higher frequencies that the standard DTMF assignments, shifting the digits and yielding a private code.

Frequency shifts work in the other direction too, as shown in Fig. 3-17. By lowering the basic rate to 1.315 MHz, an M-927 can be used to detect North American precise dial tone (350 to 440 Hz) which is decoded as * or a logic 0 on the DATA 10 pin. Similarly, a 1.315 MHz crystal to the MK-5087 yields a very close approximation of precise dial tone.

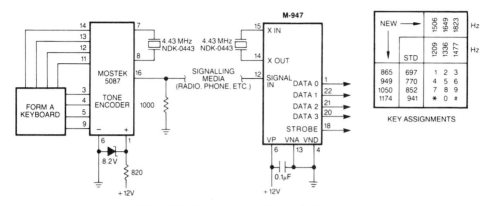

Fig. 3-16   *Custom signalling technique.*

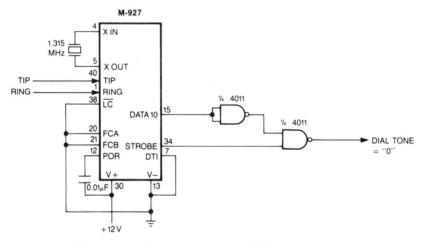

Fig. 3-17   *North American precise dial-tone detector.*

REGULATORY AGENCY REQUIREMENTS

Most countries have agencies whose purpose is to regulate the use of the public telephone network. These agencies require that certain technical specifications be met before equipment can be registered for connection to the network. In the United States, the regulatory agency is the FCC (Federal Communications Commission). Figure 3-18 shows the process used to secure registration under FCC Rules, Part 68.

Although requirements vary from country to country and from one class of equipment to another, they have many elements in common that should be considered when designing telephone equipment. If your application is regulated, consider the specific requirements in each country from the list of parameters at the end of this chapter. Actual numbers for each requirement are supplied in documentation available from British Telcom in the United Kingdom, Budespost in West Germany, AT&T or the FCC in the United States,

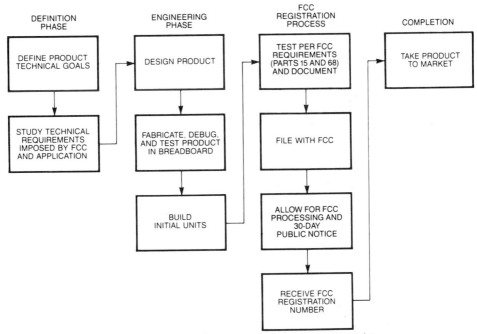

Fig. 3-18 *Producing a telephone product in the United States.*

NTTPC in Japan, and the various Ministries of Transport, Posts, or Telephone and Telegraph in many other countries.

EVALUATING DTMF RECEIVER PERFORMANCE

Deducing the performance of a DTMF receiver from its data sheet is no substitute for actual field trials or an in-depth evaluation in the laboratory. Figure 3-19 shows a test circuit similar to that used by Teltone for its own evaluations. This circuit allows the user to vary independently the level and frequencies of each DTMF signal component, to control tone burst length and interval, and to inject different types of interfacing signals.

Using this circuit, Teltone tests for errors in detection under worst-case conditions: the signal persisting for minimum duration at extremes of frequency and dynamic range. For example, the Teltone M-947 was found to have an error rate of 1 in 10,000.

PHYSICAL UNITS IN TELEPHONY

In telecommunications engineering, it is useful to express signal power as a ratio of power to some reference. The units are decibels (dB) determined by the equation:

$$dB = 10 \text{Log}^{10} \left(\frac{P}{P_r} \right)$$

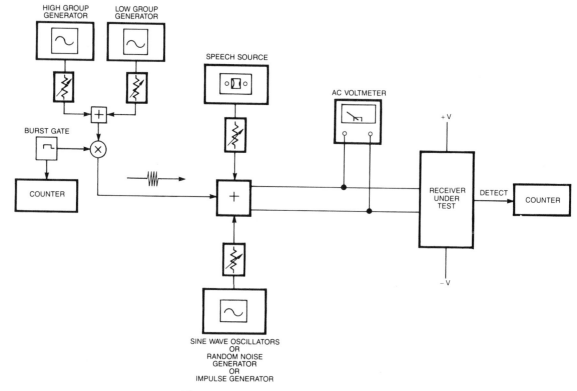

Fig. 3-19   *DTMF receiver evaluation circuit.*

where P_r is the reference power. This reference power varies from use to use and is indicated by a letter appended to dB. The most common form is dBm, where m indicates that the reference power is one milliwatt. Another common form is dBrn, which indicates that the ratio is calculated using reference noise of −90 dBm at 1000 Hz.

Because most electronic test equipment measures voltage and because voltage is the key parameter for high-impedance signal processing circuits, it is necessary to relate voltage to power and voltages to one another. Ratios of voltages are expressed in dB according to the equation:

$$dB = 20 \, \mathrm{Log}^{10} \left(\frac{V}{V_r} \right)$$

where V_r is the reference voltage. The form dBv is reserved to indicate a reference of one volt, and dBV is used for any other voltage.

When power is shown in dBm, it is customary to include a load impedance, usually 600 Ω, which allows the power to be interpreted in volts. The graph in Fig. 3-20 facilitates conversion between powers and voltages, assuming a load impedance of 600 Ω.

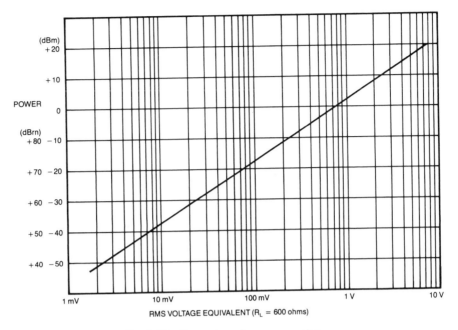

Fig. 3-20   *Power-to-voltage conversion.*

SOURCES OF ADDITIONAL INFORMATION

Teltone DTMF receivers have been used in central office, PBX, and subscriber telephone equipment for over ten years. In designing their receivers to meet the stringent signalling requirements imposed by these applications, Teltone found the following to be particularly valuable sources of information on telephony and telecommunications.

North America

AT&T Publications Catalog (PUB 10000)
Publisher's Data Center Inc.
P.O. Box C738
Pratt Street Station
Brooklyn, New York 11205

GTE Technical Interface Manual for Customer Equipment (CHB-500)
Engineering Practices and Support
GTE Automatic Electric
400 N. Wolf Road
Northlake, Illinois 60164

Private Branch Exchange Switching Equipment for Voiceband Applications (RS-46)
Electronic Industries Association
Engineering Dept.
Standard Orders
2001 Eye St., NW
Washington, DC 20006

Certificated Standard
Telecommunications Regulatory Service
Dept. of Communications
Terminal Attachment Program
300 Slater St.
Ottawa, Ontario, Canada K1A 0C8

Worldwide

The Orange Book
International Telegraph and Telephone Consultative Committee (CCITT)
International Telecommunication Union
1211 Geneva 20
Switzerland

SYSTEM PARAMETERS

Line Isolation—For safety reasons, telephone equipment must endure application of high level DC, high-level low-frequency (50-60 Hz) AC, and impulse voltages to lines connected to the public network. The application of these voltages must not cause hazardous voltages or currents to appear where they can present a danger to people or to other equipment.

Longitudinal Balance—Telephone lines connected to equipment should not have impedance differences created by the equipment which will cause common mode or longitudinal voltages to produce undesired differential voltages. This usually results in "hum" or other forms of interference from nearby sources of electrical noise. Careful balancing of line components is usually required.

Critical AC Levels—Each type of telephone equipment will probably have limits or ranges specified for AC signal levels, both generated and received, when the circuit is active.

On-Hook Impedances—To prevent interference with equipment on the same circuit and to limit loads imposed on ringing generators, the "on-hook" or idle resistances to certain AC and DC voltages usually have minimum values.

Off-Hook Impedances—To meet transmission design goals and to prevent interference with equipment on the same circuit, most regulatory agencies require that AC and DC impedances fall in specific ranges when equipment is in the "off-hook" state.

Special Frequency Assignments—Many telephone systems generate signals that can interfere with your device, and your circuit may generate signals that have a negative effect on other equipment. These frequencies are listed in the pertinent specifications.

Critical Interface Timing—Circuit events in telephone systems are usually expected at specific times. Timing for events like signal delays, pulse widths, hold times, and control intervals are detailed by regulatory agencies when necessary.

RECEIVER PARAMETERS

Because DTMF senders and receivers are designed for a unique electrical environment, the parameters used to describe their performance may be unfamiliar. Here is a brief explanation of some of the more important terms:

Sensitivity ("A" level)—The lowest level signal that is guaranteed to be detected. The range of levels detected is known as the "dynamic range," which usually extends to about 30 dB above the A level. Absolute levels are expressed in dBm, assuming a terminating impedance of 600 ohms.

Twist—The ratio of the level of the high-frequency DTMF signal component to the level of the low-frequency signal component. Most modern receivers can handle level differences of +4 to −8 dB.

Frequency Deviation—Deviations from nominal sometimes arise because of inaccuracies in DTMF senders and frequency shifts in analog carrier systems. Receivers are expected to detect signals that are off by as much as 1.5 percent plus 2 Hz from nominal throughout their dynamic range.

Signal-to-Noise Ratio—This ratio is the minimum separation between signal and noise that will not degrade detection accuracy over long strings of data. It can be specified with the signal on nominal frequency or deviated.

Detect/Reject Times—The standard choice for both signal and pause detect times is 40 milliseconds. Reject times are called for because of common telephone network occurrences that may interrupt or generate tones briefly.

Dial Tone Tolerance—Tones indicating readiness to accept dialing must not interfere with DTMF signal detection. Truly "dial tone immune" receivers tolerate dial tones more than 22 dB stronger than the A level without degrading long strings of data.

Speech Immunity—Speech and caller background noise are often present when DTMF receivers are on line. Because the human voice contains many tone combinations similar to DTMF digits, receivers must be carefully engineered to distinguish between actual signals and simulated signals. Test tapes are available that can be used to obtain typical numbers of simulations for different receivers.

TELEPHONE GLOSSARY

a wire—see line +.

b wire—see line −.

CO—central office. A local switching center to which telephones are connected by means of telephone lines. See Figure 2.

dB—decibels.

dBm—decibels above or below a reference power of 0.001 watt in a 600-ohm resistance (0 dBm equals 0.78 volts).

dial tone—a signal sent to a calling telephone indicating that digit dialing may begin. It usually consists of one or two tones between 350 Hz and 480 Hz. See precise dial tone.

DTMF—dual tone multifrequency. A signalling system used with pushbutton telephones in which the two frequencies composing each signal are taken from two mutually exclusive frequency groups of four frequencies each. See Figure 6.

end-to-end signalling—the use of DTMF signals to perform access and control operations from a remote location.

F—farads of capacitance.

FCC—Federal Communications Commission. The United States government agency which regulates the connection of equipment to the public telephone network.

hookswitch—the switch controlled by the telephone handset, which closes its contacts when the handset is lifted from its mounting and opens its contacts when the handset is replaced on its mounting.

Hz—Hertz.

line −—one of the wires constituting the voice pair of the telephone line, so called because of its connection to the negative side of the battery supply. Also called b or Ring.

line +—one of the wires constituting the voice pair of the telephone line, so called because of its connection to the positive side of the battery supply. Also called a or Tip.

max—maximum.

min—minimum.

off-hook—the condition which results when the telephone handset is lifted from its mounting, which causes the hookswitch to close its contacts and allows loop current to flow. See Figure 2.

on-hook—the condition which results when the hookswitch contacts are opened (for example, by replacing the telephone handset on its mounting) and loop current is prevented from flowing. See Figure 2.

precise dial tone—a North American standard tone consisting of 350 Hz plus 440 Hz. See dial tone.

PSTN—Public Switched Telephone Network.

pulse—a signalling system used with pushbutton or rotary dial telephones in which each digit consists of serial makes (connections) and breaks (disconnections) in the line connecting the telephone to the central office.

Ring—see line −.

rotary dialing—see pulse dialing.

telephone line—the wire loop which connects a telephone to a central office.

Tip—see line +.

toll network—the system of long-distance telephone lines available for use at a charge.

☎4

The English Style
Telephone Ringer

BEGIN YOUR FASCINATING JOURNEY THROUGH THE WORLD OF TELECOMMUNICATIONS with the assembly of the English style telephone ringer. This project is an electronic instrument designed to be an easy add-on device to just about any home telephone, a simple rotary desk or the more expensive decor telephone. This instrument, when properly connected to a telephone line, substitutes the annoying chirping-bird signaller or ringing bell into a pleasant sounding *wabble* tone. The audio output of this circuit can be compared to the pleasurable sound generated by the expensive AT&T Merlin or even the NEC or Toshiba telephone systems.

To add to the relaxing tone of an incoming call, this project incorporates the distinctive characteristics of the English type signalling system. Instead of generating the standard 2-second ring, 4-second off signal, as used in the United States, the ringer generates two short bursts of audio followed by a 4-second silence period. This gating configuration will continue until the phone is answered or the caller terminates the connection.

THE CIRCUIT

A block diagram of the high-tech telephone ringer is shown in Fig. 4-1, and the complete schematic is shown in Figs. 4-2A and 4-2B. When connected to a 12 Vdc wall transformer (T1), IC1 (4069, which generates three different audio signals) goes into oscillation. Hex inverters IC1a and IC1b form a low-frequency generator. This section of the 4069 oscillates at 12 Hz, IC1c and produce a 287 Hz signal, and IC1e and f produce a 335 Hz signal. Whether the telephone is being used or not, IC1 will effortlessly continue to create the three needed signals.

The secret of producing the pleasant sounding audio tones is the way the three base frequencies are digitally mixed by the 4093 IC (quad 2-input NAND Schmitt trigger). Once combined, the now continuous wabble tone is present at pin 11 of IC2c.

To shape the tone ringer output into the desired form, the 4066 IC is used. Three of the four bilateral switches of the IC are wired in series. See Fig. 4-3 for

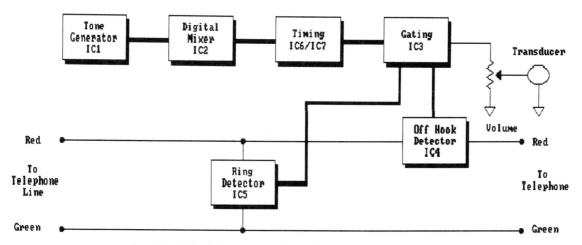

Fig. 4-1   *A block diagram for the English style telephone ringer.*

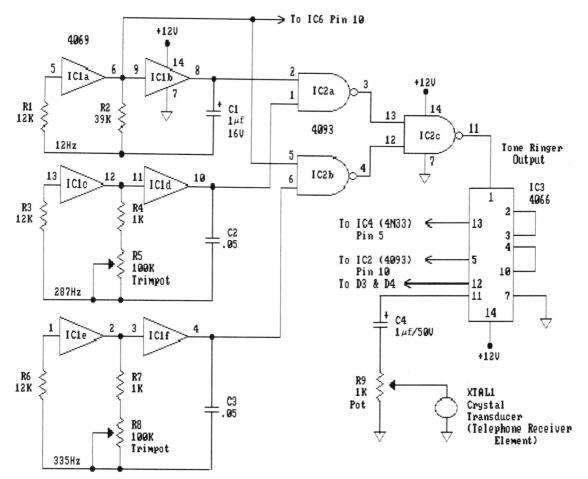

Fig. 4-2A   *A partial schematic of the telephone ringer showing the tone generators.*

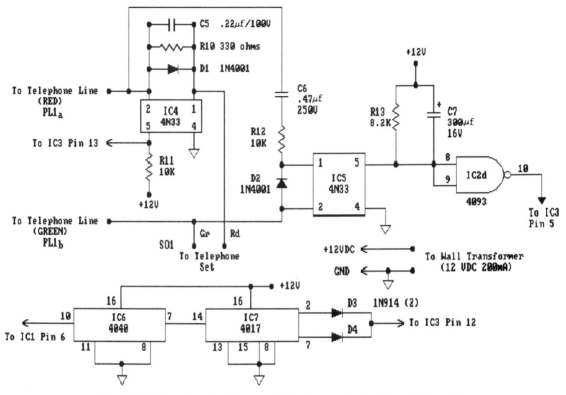

Fig. 4-2B   *The schematic diagram of the tone gating for the telephone ringer.*

a block diagram illustrating the internal connections of this integrated circuit. With the audio tones from pin 11 of IC2c connected to pin 1 of IC3, it is easy to shape and control the ringer output to the crystal transducer (XTAL1).

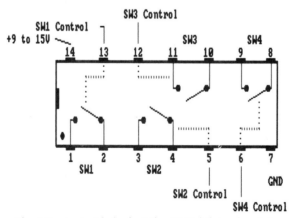

Fig. 4-3   *An inside look at the 4066 bilateral switch.*

CONTROL CIRCUITRY

Creating the English Type Telephone Ringing Signal

To formulate the English type ringing signal, a low-frequency clock is needed to generate a series of positive-going pulses. Diodes extract the desired pulses capable of producing the control signal needed at pin 12 of IC3. The perfect clock generator is located at pin 6 of IC1. To refresh your memory, this clock generates 12 pulses every second. You cannot use this frequency directly because this 12 pps (pulses per second) is still too high. To create a more desirable timing output, a CMOS IC 4040 (IC6—12-Stage binary ripple counter) is used. This IC divides the 12 pps clock into a more suitable 3 pps output. This 3 pps signal can be seen with an oscilloscope at pin 7 (IC6).

Now that you have a clock signal of 3 pps, you need an IC that can convert this clock into positive-going SINGLE voltage outputs. The 4017 IC (IC7) would be the perfect choice for this application. This CMOS IC is a divide-by-10 counter with a 1-of-10 positive output (for every clock pulse delivered to its input, the 4017 will deliver one positive pulse at one of its 10 output pins). When the tenth pulse is received, the process starts again.

By connecting two diodes (D3 and D4), one to pin 2 and one to pin 7, of IC7, you can simulate the English telephone ringer by allowing the first clock pulse to deliver a positive-going pulse to the control lead (pin 12) of one of the solid-state switches located within the body of the 4066 IC (IC3). This positive pulse will instruct the 4066 to close the switch located between pins 10 and 11. Whatever signal is located on its input pin (pin 10) will be allowed to flow to its output (pin 11).

For illustration, assume that the two additional internal switches located between pins 1 and 2 and pins 3 and 4 of the 4066 are not connected. With the second clock pulse delivered to the input of IC7, the 4017 now returns pin 2 (which had the positive control signal) to ground. With the control signal now at pin 4, the signal that was once delivered to pin 12 of the 4066 is now gone. Without this pulse, the wabble tone will be prevented from passing to the output of the switch. This will occur only until the input of the third clock pulse. This pulse will now create a positive signal at pin 7 of IC7. The voltage is now delivered through D4 (1N914) to the same control lead of the 4066. The internal switch is instructed to close until the fourth pulse is received. Seeing that no other connections are made to the control lead of the 4066, the fourth, fifth, sixth, . . . , ninth, and tenth pulse will generate NO audio output. In effect, the tone ringer is turned off for the duration of these clock pulses. This is true until the eleventh pulse is received. At this time, IC7 will recycle back to its starting position, where a positive-going pulse is delivered to pin 2 of the IC. Here the switching process repeats continuously.

Figure 4-4 illustrates the gating pulse that takes place during this timing process. Note, that with every peak shown in this diagram, a wabble tone has been delivered to the output of the 4066 IC. This pulse diagram illustrates the English Style Telephone Ringer.

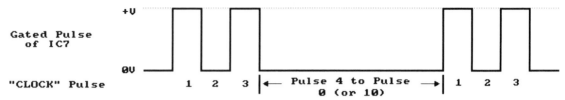

Fig. 4-4 *An oscilloscope display of the gating pulse if it were connected to the D3 and D4 junction (IC7 output).*

Ring Detection

A second control signal must be derived that will deliver a positive pulse to IC3 (pin 5) every time an incoming call is being received. This is easily accomplished by using an opto-isolator IC (IC5). This IC not only will detect a ringing signal on a telephone line but will maintain maximum isolation between the line and the circuitry of the telephone ringer. The isolation is a must according to FCC rules and regulations regarding customer-installed telephone apparatus.

The voltage applied by the telephone company central office to activate a signalling device when a call is being received is about 90 Vdc with a frequency of 20 or 30 Hz. This voltage is applied to the small internal LED (light-emitting diode) of IC5 (pins 1 and 2). Capacitor C6 restricts the flow to dc voltage. If this capacitor were not present, the dc voltage on the telephone line would keep the internal LED of the opto isolator on all of the time. This would be of no use.

To get back to the ring detector, resistor R12 (10 kΩ) lowers the 90 V ring signal to a value that can be easily handled by the internal LED of the 4N33 IC. Across pins 4 and 5 of the same IC is a light-sensitive transistor. This transistor will conduct a current when its associated LED begins to flash. This conducting grounds the voltage that was once present at pin 5. This grounding signal is then applied to pins 8 and 9 of IC2. The 4093 which is used at this point is a quad two-input NAND gate using a Schmitt trigger construction.

The Schmitt trigger is used at this point because the snap-action voltage delivered by this device is needed to prevent the slow discharge voltage of capacitor C7 from entering the 4066 IC. This discharging will adversely affect the operation of the circuit by allowing the tone output to be loud when the capacitor is first charged. Then the tone output slowly decreases in volume during the normal off cycle of the incoming ring signal. With the problems generated by using C7, it is still needed. It will keep the output of IC5 low while the ring signal across the telephone line is at its 3-second off cycle. In other words, capacitor C7 will produce (with the help of the Schmitt trigger NAND gate), a positive-going control signal from the time the first incoming ring signal is detected until the call is answered or terminated. This signal is then applied to pin 5 of IC3 (4066).

At this time, you have an English style ring being controlled at IC3. You also have a positive voltage being generated every time a telephone call is being received. With these two control pulses, you have a very effective telephone ringer with one small problem. When a call is being received, capacitor C7 charges up. This charging action develops a control signal that is being applied to the 4066 IC. At this time, say you lift the handset of the telephone. The tone ringer will still produce an audio signal until C7 discharges sufficiently to overcome the snap-action of the Schmitt trigger NAND gate (IC2). This discharging process takes about 4 to 5 seconds.

To overcome the shutoff delay, an off-hook detector is used. This detector is made from a second 4N33 opto-isolator IC. When the telephone is off-hook, the internal LED of the IC lights, causing the associated transistor to conduct. This conduction will cause pin 13 of IC3 (4066) to go to ground. When this control lead is grounded by lifting the telephone handset, the electronic switch located between pins 1 and 2 will cease to conduct, thus literally disconnecting the tone ringer from the transducer (XTAL1).

When the handset is placed on its cradle, the LED across pins 1 and 2 of IC4 will extinguish. This will prevent any light from falling on the light-sensitive transistor, thus preventing conductivity. At this time the ground, which was once located at pin 5, is now a positive voltage. With a positive signal at this point, the control lead of IC3 pin 13 will instruct the IC to close its associated switch.

With the Switch across pins 1 and 2 of the 4066 IC closed, all that remains is to have a positive signal applied to pin 5 of the 4066 from the ring detector (IC5). This positive signal will allow the tone ringer to operate.

POWER REQUIREMENTS

The entire English telephone ringer does not require a fully regulated power supply. So to save precious PC board space, an on-board rectifier and filter was omitted from the design. Seeing that the telephone ringer requires only 100 mA for operation, you might want to consider using a 12 Vdc, 200 mA wall transformer. It is highly recommended that this type of power supply be used. Make sure that the transformer you purchase is a UL (Underwriters Laboratory) approved type. These approved transformers have been rigorously tested, and they provide high isolation between the user and the deadly household current.

CONSTRUCTION

To guarantee a working project when power is first applied, be sure to follow these directions carefully. Careful assembly now will keep you from wasting time troubleshooting the ringer later. The telephone ringer can be wired on a standard perforated board, but care must be taken in its use. The FCC requires that adequate isolation be maintained between telephone lines and power supply voltages. For this reason, a printed circuit board is highly recommended. The artwork for the recommended PC board is shown in Fig. 4-5. The compo-

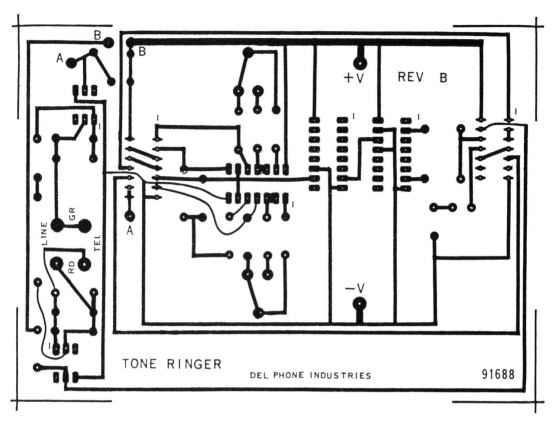

Fig. 4-5 *The PC board artwork for the telephone ringer shown at a scale of 1 to 1.*

nents parts layout is in Fig. 4-6. The illustrations can be used to fabricate your own board, or you can send it out to a circuit board manufacturer. The choice is up to you. Both methods will produce a board of high quality. If you wish, you can buy a fully etched and drilled board from Del-Phone Industries of Spring Hill, Florida. Drop them a line and request current prices on this and other telcom kits and components.

Whether you wire the telephone ringer on a perf (perforated) board or the recommended printed circuit board, start the assembly by installing the fixed-value resistors. All resistors listed for use in the Telephone Ringer are ¹/₄ W, 5 percent resistors. All capacitors in the parts list are listed in μF microfarads. Next, install the capacitors, making sure of proper orientation for all electrolytes. Mount the four diodes next. Just remember that one pair of diodes are 1N4001, and the second pair are glass diodes (1N914). Again, make sure of proper orientation and component number.

The last items to be installed are the seven ICs. ICs are heat-sensitive components, so be sure you use an adequate heat sink when soldering. Also take note of the orientation of all ICs. An improperly installed IC can be damaged in no time at all. So be careful.

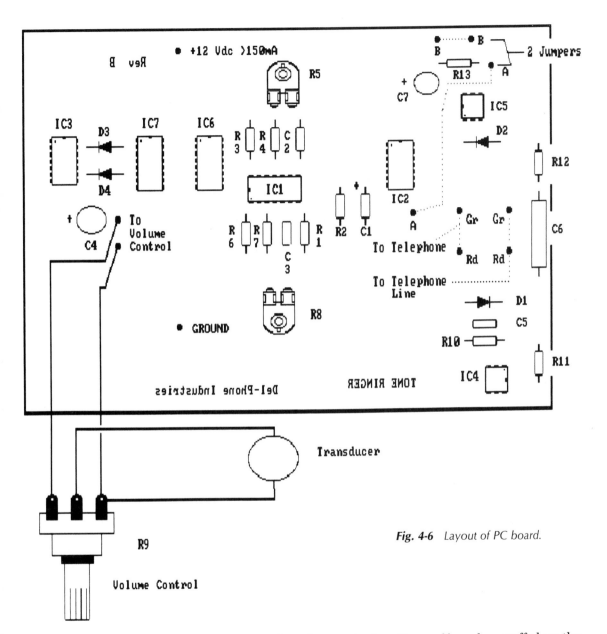

Fig. 4-6 *Layout of PC board.*

When all components have been wired on the perf board or stuffed on the PCB (printed circuit board), you can now install the volume control (R9). This potentiometer is a back-panel control that should be easily accessed at will.

The crystal transducer (XTAL1), which can be removed from an old telephone handset, is wired directly to the volume control as shown in the schematic. When wired, the transducer should be glued to the top of the desired housing. Drill a number of holes in the cabinet before installing the crystal to allow the audio this is produced to easily penetrate the air.

When you have all parts installed, check for the standard solder and wire bridges, proper orientation of all ICs, capacitors, and diodes. When inspection is complete, connect the power leads from the dc wall transformer (T1—12 Vdc 100 mA). Seeing that this transformer delivers a dc instead of an ac voltage, you must determine the polarity of its leads before soldering T1 to the circuit. In most cases, the positive lead of the transformer will be marked with a white tracer.

When T1 polarity has been determined and its leads soldered to their proper location, plug the transformer into a standard ac outlet. Connect the black lead of a voltmeter to the common ground connection of the telephone ringer. Connect the meter red lead to the power pins of each of the seven ICs. The meter reading for all ICs should be 12 Vdc. If you have a voltage lower than 12 V, remove power and check for improperly installed ICs and solder bridges. Solder bridges often show up between the pins of the integrated circuits.

CIRCUIT TESTING

To provide an audio output from your telephone ringer (for testing and adjustments), place a temporary short between pins 4 and 5 of IC5 (4N33 IC). Some sort of tone will be emitted from the crystal transducer (XTAL1).

If a frequency counter is available, place the black lead of the meter to the common ground connection of the ringer. The counter red lead should be placed first at pin 10 of IC1. At this location, the hex inverters of the 4069 should be generating a frequency of 287 Hz. If not, use a small screwdriver to adjust the 100 kΩ trimmer potentiometer (R5) until you get such a reading.

When complete, move the red lead of the frequency counter to pin 4 of IC1. This section of the hex inverter should be generating a frequency of 335 Hz. If not, adjust R8 until this reading is indicated on the meter. The third oscillator (comprising IC1a and IC1b) is fixed at 12 Hz and cannot be adjusted.

If no frequency counter is available, you can try to adjust by ear. Then adjust both 100 kΩ trimmer potentiometers (R5 and R8) until the most pleasant tone is created by the transducer. Remove the short you have intentionally installed between pins 4 and 5 of IC5.

TELEPHONE MODIFICATION

With your English style telephone ringer now in operation, it is time to consider which telephone you will connect to the ringer. This telephone can be either a rotary or tone-dial instrument, a telephone manufactured by ITT, or an inexpensive import. The only internal wiring that must be made is to an ITT telephone or similar make. ITT telephones contain a single- or double-gong bell, and imports contain an inexpensive tone type ringer. The modification explained in this section will prevent the phone itself from ringing when an incoming call is being received.

For the owners of an imported telephone, the manufacturers of these instruments include an on/off switch for the ringer. All you must do is to flip this switch to the off position. The import is now ready to be connected to the telephone ringer. If you have an AT&T or ITT telephone, read on.

Now, consider the more expensive AT&T or ITT telephones. This modification is more involved than just flipping a switch, so read the following very carefully. The first thing to do is to remove the housing from the phone. If the telephone is a desk type, just turn the phone over and locate and remove the two screws that hold the housing to the base plate.

With the housing removed, locate the mechanical telephone ringer. This ringer might have either two or four wires connected to the bell coil. For a four-wire ringer, the colors are red, black, slate/red, slate/black (see Fig. 4-7). For a two-wire bell, (a 148BA ringer is usually found in a 200 or 2200 desk and wall phone and in the 2554 phone) the colors are red and black (see Fig. 4-8). Do not confuse the bell red wire for the red wire of the telephone line cord connection. They are two completely different things.

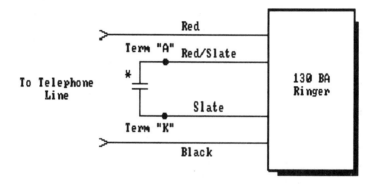

NOTE: This capacitor is located inside the telephone network between Terminals "A" and "K"

Fig. 4-7 *The wire connections from a 130BA ringer to the telephone line. Note that the capacitor is placed in series with the ringer coils.*

Whether you have a two- or four-wire telephone ringer, just remove any one of the bell leads. (Make note of the location you remove the wire from. You might want to rewire the phone back to its normal configuration.) With one of the wires now removed, take a strip of electrical tape and cover the lead. This taping prevents short circuits inside the phone once the housing is replaced.

WIRING THE TONE RINGER TO THE TELEPHONE

Now that you have a telephone with a disabled ringer, it is time to wire the unit to the telephone ringer. With the line cord from the telephone you wish to use, connect the red wire to PL1s red connection point and the green to its green connection point (see Fig. 4-2). If you have wired the telephone ringer on

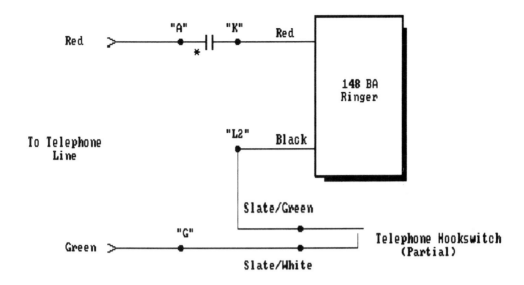

NOTE: Capacitor located inside the telephone
network between Terminals "A" and "K"

"X" Denotes Network Terminals

Fig. 4-8 *The wire connections from the 148BA ringer to the telephone line.
Note that the capacitor is placed in series with the ringer coil.*

a perforated board, solder the red lead to pin 1 of IC4. Solder the green wire of
the line cord to pin 2 of IC5.

If your telephone uses the newer modular line cords, you can purchase a
plastic mating connector from any Radio Shack store. With this mating connector in hand, just solder the red and green wires to the appropriate locations as
mentioned above and as shown in Fig. 4-2. To connect your phone to the telephone ringer, just take the telephone modular line cord and snap it into its mating connector. All electrical contacts are made within the connector.

You might have noticed that your particular telephone may have two additional wires in the line cord. The colors are yellow and black. Normally these
wires are not used, so just use a piece of electrical tape and cover them up to
prevent any unwanted shorts in the system.

If you install your finished telephone ringer with an AT&T or ITT trim-style
wall telephone, take care when reinstalling the housing to the base of the
device. You must hold the hook-switch plunger up, as shown in Fig. 4-9, to
avoid damaging the hook-switch. Other than that, there are no special considerations that must be taken during installation.

CONNECTING THE TELEPHONE RINGER TO THE LINE

All that is left now is to wire the ringer to the incoming telephone dial tone
line. Figure 4-2 illustrates these connection points as SO1 "To Telephone Set."

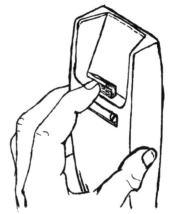

Fig. 4-9 *With a Trendline wall telephone, take care in the removal and replacement of the base housing. Do not damage the hook-switch.*

Take the red wire (from the telephone line) and connect this lead to pin 2 of IC4. Connect the green wire to pin 2 of IC5. Just remember to maintain proper polarity throughout the system. In other words, connect all red wires to red and all green wires to green (see Fig. 4-10).

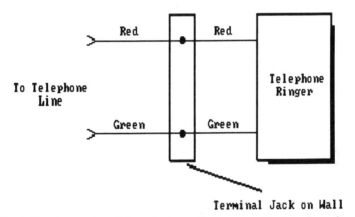

Fig. 4-10 *How to install the telephone ringer across the telephone lines.*

If your home uses new modular wall jacks, Radio Shack also stocks telephone line cords with a mating connector on one end. Just purchase this cord and connect the red and green wires as indicated above. When the wires are connected, just plug the mating connector into the modular wall jack.

If you wish to use the ringer on a multiple telephone line, refer to Fig. 4-11. This illustration shows that the finished ringer is wired from the common line connecting all phones (which are connected in parallel), to the incoming dial tone line. Just remember to observe line polarity.

TESTING THE ENGLISH STYLE TELEPHONE RINGER ON LINE

With a single-line desk or wall telephone connected to the ringer, have a friend give you a call. When the call comes in, the telephone ringer will emit a

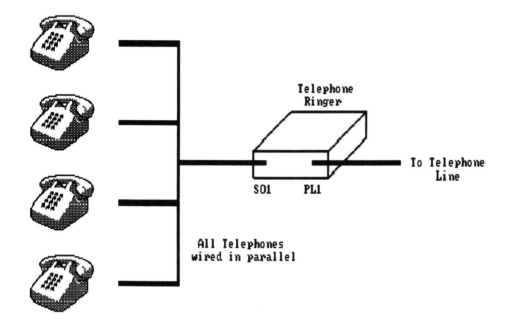

Shared Telephone Ringer System

Fig. 4-11 *How to connect the telephone ringer in a multiple telephone installation.*

wabble tone from the crystal transducer (XTAL1). Adjust R9 for a comfortable volume level. (See Fig. 4-12 for a label for volume adjustment.)

Lift the telephone receiver; immediately the telephone ringer will cease to operate. At this time, normal conversations can take place. When the call is complete, just return the handset to its cradle. The telephone ringer will automatically rearm itself, waiting for its next incoming ring signal.

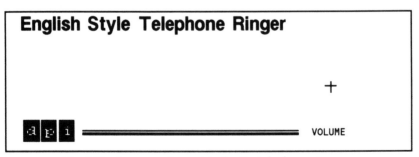

Fig. 4-12 *Front-panel template for the telephone ringer.*

PARTS LIST FOR THE ENGLISH STYLE TELEPHONE RINGER

R1	R3	R6	12 kΩ resistors
R2			39 kΩ resistor
R4	R7		1 kΩ resistor

R5 R8	100 kΩ trimpots (Digi-Key K4A15)
R9	1 kΩ audio taper potentiometer
R10	330 Ω resistor
R11 R12	10K Ω resistors
R13	8.2K Ω resistor
C1	1 μf/16 V electrolytic capacitor (radial)
C2 C3	0.05 μf disk capacitor
C4	1 μf/50 V electrolytic capacitor (radial)
C5	0.22 μf/100 V disk capacitor
C6	0.47 μf/250 V tubular capacitor (axial)
C7	330 μf/16 V electrolytic capacitor (radial)
D1 D2	1N4001 diode
D3 D4	1N914 diode
IC1	4069 hex inverter
IC2	4093 quad dual-input NAND Schmitt trigger
IC3	4066 quad bilateral switch
IC4 IC5	4N33 opto-isolator
IC6	4040 12-stage binary ripple counter
IC7	4017 divide-by-10 counter with 1-of-10 outputs
XTAL1	High-Z crystal transducer (telephone type crystal element—available from Del-Phone Industries[*])
T1	12 Vdc 200 mA wall transformer
1	telephone—either rotary or tone dial
1	housing
1	telephone line cord (standard or modular)
1	printed circuit board (available from Del-Phone Industries[*])

---

[*]Del-Phone Industries, P.O. Box 5835, Spring Hill, Florida 34606

☎5
The Two-Tone Ringer

THE SECOND TELEPHONE RINGER PRESENTED HERE IS AN UNUSUAL TWO-TONE SYSTEM designed to generate a distinctive twin-note alarm signal instead of the standard ringing of a bell. This unique design, using two 555 ICs, will allow the output of IC1 to control the frequency of IC2. With the part values shown in the parts list, the first 555 (IC1) operates as an astable multivibrator which runs at about 3 Hz. IC2, which is also an astable multivibrator, functions at a center frequency of about 1.5 kHz. By allowing this 1.5 kHz tone to be shifted in frequency six times a second by the 3 Hz circuit (IC1), a distinctive tonal output is produced.

By incorporating a home-brewed ring detector and a 4066 IC, you have the makings of a very remarkable telephone ringing system.

HOW IT WORKS

The pinout of the two-tone telephone ringer is shown in Fig. 5-1, and the schematic is shown in Fig. 5-2. With the values provided, IC1 will generate the 3 Hz signal. This signal is accomplished by the relatively large C2 and R5 values. When power is first applied, C2 is charged through the series connection of R4 and R5. Depending on these values, C2 can be charged relatively fast or extremely slow. Once charged to its peak value, the 555s *threshold* comparator senses this and flips the internal circuitry to the opposite state. In other words, when the 555 senses that C2 charged to about two-thirds of the supply voltage, it flips a logic 1 output to its compliment logic 0.

Then, when at logic 0, C2 starts to discharge through R5, until the voltage drops to about one-third of the supply voltage. At this point, the internal trigger, once again senses this voltage and flips the circuit back into its initial state (0 back to 1). The frequency of this flip/flop action is dependent on the values of R4, R5, and C2.

To generate higher frequencies, such as those created by IC2, the component values of R8 and C3 have been reduced. By reducing the series resistance values, capacitor C3 can charge and discharge faster, thus creating a faster flip/flop action at the output (pin 3).

By connecting the output of IC1 to the junction of resistors R7 and R8 of IC2, the frequency of the signal supplied to pin 3 (of IC2) will be varied at a rate of six times per second, thus creating the distinctive dual-tone signal.

GND	1	8	Positive Supply Voltage
Trigger	2	7	Discharge
Out	3	6	Threshold
Reset	4	5	FM Input (Optional)

Fig. 5-1 *Pin assignments on the widely used 555 timer/multivibrator IC.*

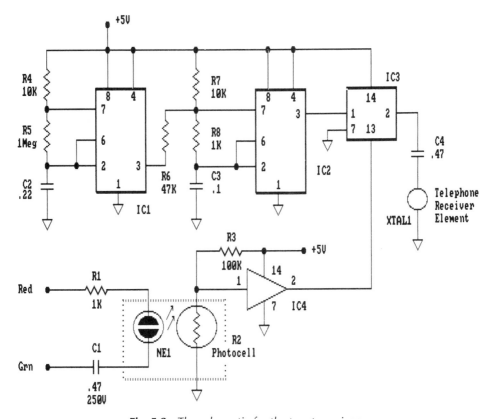

Fig. 5-2 *The schematic for the two-tone ringer.*

By using a conventional 4066 (bilateral switch), the audio output of the dual-tone generator (IC1 and IC2) can be controlled by placing either a ground or logic 1 at the associated control pin of the switch (IC3). If this control pin were to have a + 5 V or 1 applied, the two-tone output of IC will be allowed to pass through to a high-impedance telephone receiver element where the audio of the dual-tone is reproduced. If a ground or 0 were applied, the internal switching action of the 4066 will prevent any sound from passing through, thus turning off the ringer. Of course, this control or *gating* period will be created by the presence or absence of the ringing signal across the telephone line. This is the job of our home-brewed ring detector (NE1 and R2).

You know by now that the telephone company sends out a 90 V 30 Hz signal to indicate an incoming call to your telephone. You can take this 90 V and have it flash a neon bulb (NE1 is part of the ring-detection circuitry—see Fig. 5-2). When this bulb flashes, the resistance of the photocell (R2) drops, thus completing a ground path for the voltage across resistor R3. By applying this ground pulse to IC4 (Schmitt trigger plus invertor), it is first cleaned up and made into a pulse usable with digital electronics (this type of signal contains no extraneous noise or pulse information). And, the pulse is inverted into a positive going signal which can now be applied to the control pin of the 4066 switch (IC3).

In the absence of a ring signal, NE1 ceases to flash, thus allowing the resistance of the photocell to increase. This increase of resistance allows R2 to act pretty much as an open circuit. The open circuit created by R2 allows the full 5 V to be applied to the input of IC4. The output, which is inverted, applies a 0 to the control pin of the 4066. With a 0 at pin 13, the tones generated by IC1 and IC2 will not be allowed to pass through, and the receiver is silent. To understand the on and off pulses of this ring detection a little better, please refer to Fig. 5-3.

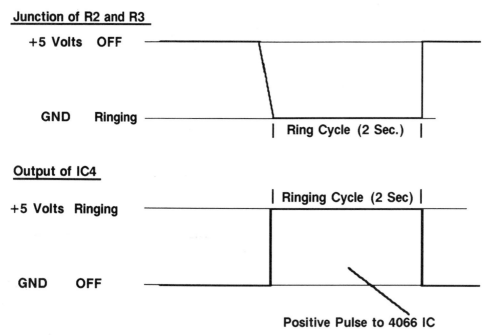

Fig. 5-3 *The incoming ringing signal as seen by the photocell (R2).*

ASSEMBLING THE RING DETECTOR

The ring detector used with this project is a homemade component that you can assemble in no time at all. All you need is a standard neon light bulb and a photocell, both of which can be purchased at your local Radio Shack store. By referring to Fig. 5-4, you can see the detailed assembly needed to build the detector.

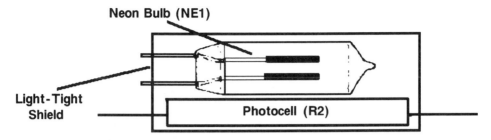

Fig. 5-4 *Do you have a neon bulb and a photocell lying around?*
If so, you can make a ring detector yourself and save a couple of bucks to boot.

To assemble the detector, just place the bulb and photocell in physical contact with each other. Then take a length of black electrical tape and cover this component sandwich. The taping of these components must provide a light-tight enclosure (remember that the leads of the components must be allowed to protrude from the electrical tape). That is all there is to it.

CONSTRUCTION OF THE RINGER

The assembly of the two-tone telephone ringer does not require any exotic components or test equipment. Probably, you have the makings of the ringer already in your electronic junk box. If the junk box theory is your avenue of construction, just remember to test the parts before you install them in the circuit. Speaking from experience, junk box parts or even factory rejected components that have been purchased can transform an hour's work of assembly time into an ordeal of three or four hours of troubleshooting and reassembly. So it is wise to test all so-called bargain components.

Once you have all the parts, the easiest and the most economical way to a finished project is to use a perf board. The term *perforated* or *perf* board for short, is used to describe a sheet of nonconductive material. The material has been bombarded with holes every 0.1 inch. Because of its nonconductive nature, the perf board makes an ideal base on which to build electronic projects. For a price of $7.00 for a 4 × 17 inch size, the perf board is the economical savior to the shoestring electronic investor.

Now that you have the parts and a piece of perf board, get down to some electronic assembly. For the first-time builder, use 8-pin IC sockets for both IC1 and IC2 and two 14-pin sockets for IC3 and IC4. This will, without a doubt, increase the overall cost of the project, but you will not be confronted with the grim task of rewiring the ringer just to replace burned-out ICs.

Because the ringer is such an easy assembly job, there really is not much to say about its construction. The only components that might be placed in backwards are the four integrated circuits. All the other parts do not have polarities, so inserting them haphazardly will not cause you to lose any sleep. So when it comes to soldering or inserting the ICs in place, be sure to get the proper orientation.

When you have completed the assembly of the ringer, inspect your workmanship for solder bridges and especially mis-wires. Another thing to watch out for, especially if you wire the circuit on a perf board, are the little gremlins that take the form of tiny bits of bare wire. They have the reputation of getting stuck in the most destructive points of a circuit. So when inspecting the finished board, keep a sharp look out for those little gremlins. They just love to "pop" a newly finished project.

Have you finished your inspection? Did not find any demented demons in the wiring, did you? That is good. Apply the +5 V to the circuits and let this baby wail.

The cheapest way to build a 5 V power supply for every project is to consider purchasing a dc wall transformer. These transformers are UL-approved devices that plug directly into a standard ac socket. The internal step-down transformer delivers a safe, low-voltage output that is rectified and filtered into a fairly good dc signal source. The only consideration you must heed is the current rating of the unit. This project gobbles up about 100 mA of current. So a dc wall transformer proudly displaying a current rating of 150 mA or higher will work just fine in this case. Remember that a circuit will only eat enough current to make it operate (unless there is a wiring error). So if you use a power supply delivering a higher amount of current, the operation of the circuit will not be hindered in any way.

TESTING AND INSTALLATION

So with +5 V applied to the proper locations on your board, temporally place a short from pin 1 of IC4 to ground. By placing this pin at ground, you can simulate a ringing signal across the telephone line. With this short applied, pin 13 of the 4066 will be high (logic 1). This voltage will allow the generated dual-tone signal being created by the two 555s to be heard at XTAL1 (telephone receiver element). If no tone is heard, recheck the wiring of both 555s and the 4066 switch. If you must, you can use the telephone receiver element (with a 0.47 µF capacitor in series) as a test probe. First place the receiver (with one side at ground) to pin 3 of IC1. You should hear a rapid pulsing. This would be normal. Next, place the receiver at pin 3 of IC2. At this point, there should be a dual-tone output. Then place the receiver at pin 1, then at pin 2 of IC3. Located at pin 1, you should be greeted with the now-familiar two-tone signal, but if pin 13 contains a low (logic 0), there will be no output at pin 2 of the 4066. If you wish, disconnect pin 2 of IC4 and place your own control pulse at pin 13. Just take a piece of wire and solder it to pin 13 of IC3. Take the other side and place it on the +5 V input feed. If all is well up to this point, the dual-tone signal will be penetrating the air. If not, your problem lies within the circuitry of IC4 and its associated components. So good luck with your troubleshooting.

For now, get back to the testing of your telephone ringer. With a short applied from pin 1 to ground of IC4, you will hear the signalling tone. When removing the short, the signal will stop. So far, so good.

Now connect the circuit across the telephone line. To do this, just take the leads protruding from the neon bulb components (R1 and C1) and connect them to the red and green wires located within the junction box screwed to the baseboard of your house or apartment. Get a friend to call you. At that time, your main telephone will ring and so will the two-tone telephone ringer. This test illustrates that your project is operational and that you can now disconnect the irritating bell of your main telephone.

DISCONNECTING THE BELL

You now have a working dual-tone telephone ringer. Just like the English style ringer, the standard telephone bell must be disconnected from the line. To do this requires no great amount of time nor tools. As with the English telephone ringer, all you must do is to remove one of the four (or two) wires that are connected to the bell. Tape it and store it for future re-connection if you ever decide you would rather have the annoying bell. For a more complete explanation of this telephone rewiring, please flip back to chapter 4.

For any telephone ringer project you might build, whether it is a book design or your own, the procedure for the disconnection of the telephone bell is the same for all circuits. Just remember to tape the dangling bell wire, or you run the risk of shorting the incoming ring signal to ground. Undoubtedly, the telephone company would not greatly appreciate the short.

FINAL ASSEMBLY

Once your ringer is operating without any problems, turn your thoughts to housing the circuit. Plastic cabinets are an excellent choice for housing projects. They are cheap and easy to drill in comparison to sheet metal panels. Also, they can be purchased from just about any electronic mail order house. Radio Shack has a variety of housings to choose from. Once you have purchased a cabinet and drilled the needed holes for the telephone line and power cord, you might want to use the template illustrated in Fig. 5-5. This front-panel drawing can easily be copied by any copy machine. Then glue the copy in place. Templates of this type can give your finished project a look that will be the envy of electronic technicians.

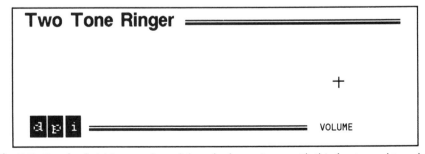

Fig. 5-5 *Put the finishing touches of the telephone ringer with this front-panel template.*

PARTS LIST FOR THE TWO-TONE TELEPHONE RINGER

R1	R8	1 kΩ resistor
R2		standard photocell (Radio Shack model)
R3		100 kΩ resistor
R4	R7	10 kΩ resistor
R5		1 MΩ resistor
R6		47 kΩ resistor
C1		0.47 μF 250V capacitor
C2		0.22 μF 50V disk capacitor
C3		0.1 μF 50V capacitor
C4		0.47 μF 100 V tubular capacitor
IC1	IC2	555 timer ICs
IC3		4066 Bilateral switch IC
IC4		4584 Schmitt trigger/inverter IC
XTAL1		Telephone receiver element (or equivalent)
NE1		Standard neon bulb
1		5 V 150 mA dc wall transformer
1		Perforated board or printed circuit board
1		2- or 4-conductor telephone line cord
1		Project housing of your choice

☎6

Electronic
Telephone Ringer

CHAPTER 4 PRESENTED A CIRCUIT THAT SIMULATES THE DISTINCTIVE TWO-RING SIGNAL of an English style telephone. For a newcomer to the world of electronics, this project might seem to be too intricate. This perception is caused by the extensive use of signal generators, mixers, and dividers. This digital signal generation might be considered to be involved. But take heart. If you desire a circuit that will simulate the English telephone ring but do not have the extensive knowledge of digital electronics, this chapter is for you.

By using a circuit called a half-monostable multivibrator, you can complete a design that distinguishes the starting and finishing pulse of a telephone ring signal. Also classified as *edge-triggering* circuits, inverters wired as shown in Fig. 6-1 can yield an output pulse on either the leading or trailing edge of an input signal. The time constant or width of the output pulse is determined by a number of factors. The two most critical factors are the values of the charging resistor(s) and the value of the series capacitor. Another factor that plays a role in determining the width of the output pulse is the internal resistance of the IC itself.

Take a closer look at this most unique electronic building block. Figure 6-1A illustrates an incredibly simple and easy to construct, half-monostable circuit. With each rising edge of an inputted pulse, capacitor C charges. By using a B-series or Schmitt trigger CMOS IC, the device turns on. With the input pulse on its positive-going stretch, capacitor C begins to discharge. Once the capacitor discharges to a certain level, the internal snap action of the Schmitt trigger, turns off the output voltage thus creating a fast on/off pulse only when an input signal goes from a ground to a positive voltage.

In contrast, take a close look at Fig. 6-1B. This circuit detects the trailing edge of input pulses. By applying a voltage on one side of the charging capacitor (C) through resistor Ra, you can bring this voltage abruptly to ground (or as close to ground as resistor Rb will allow), by introducing the positive-to-ground edge of a triggering pulse. Once the input pulse is at ground potential, the Schmitt trigger turns the IC on, thus allowing a voltage at its output pin. During this turn on period, capacitor C recharges from its now ground state to a

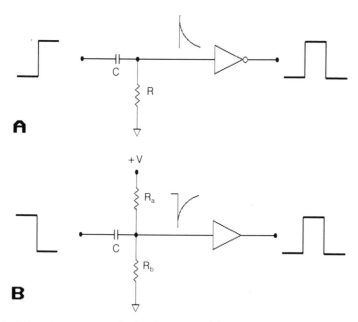

Fig. 6-1 *Triggering a gate at the leading (A) and the "Trailing" edge (B) of a pulse.*

positive voltage. When the trip point level of the Schmitt trigger is met, the IC will again turn off, thus providing a short positive going pulse from a trailing edge of an input signal.

Now that you have an idea on how half-monostable multivibrators operate, consider what you have learned and apply these thoughts to an incoming telephone ring signal.

HOW IT WORKS

For now, turn your attention to the schematic of the electronic telephone ringer (Fig. 6-2). As usual, you have a ring detector comprising resistor R1, capacitor C1, diode D1, and the now-acclaimed opto-isolator 4N33.

When a telephone ring signal is introduced to the line by the central office, the internal LED of the 4N33 begins to flash, thus allowing the Darlington transistor array located across pins 4 and 5 to conduct. When this array conducts, the voltage present at pin 5 is now shunted to ground. During the off period of the incoming ring cycle, this voltage once again goes high.

At this point you need positive-going ring pulses as your input to the half-monostable circuits. You need to invert this signal. To do this, use a 4584 hex inverter IC (IC2). This inverter is a Schmitt trigger inverter, and it must provide two distinct functions. The first function is that it must invert the signal present at pin 5 of IC1. The second is that it must provide a buffer action between IC1 and the two half-monostable circuits (IC2 b and c).

For a better understanding of the next stage of signal development, please refer to Fig. 6-3. Shown in Fig. 6-3A is the now-inverted ring signal as created

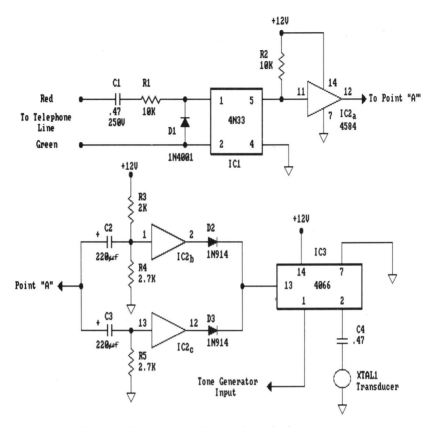

Fig. 6-2   *The schematic for another telephone ringer.*

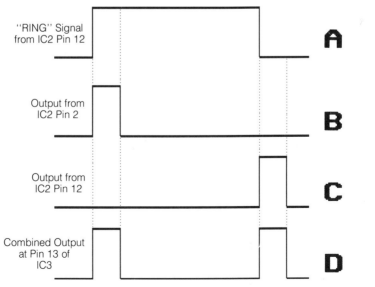

Fig. 6-3   *Circuit pulses and gating signals at various points in the telephone ringer.*

by the incoming voltage across the telephone line. It is this pulse that you will cut apart to create two distinctively different timing pulses.

From the schematic in Fig. 6-2, the output from IC2a (pin 12) is delivered to capacitors C2 and C3. These capacitors are relatively large because you need to generate two pulses, both lasting about one-quarter of a second. Components IC2b and IC2c make up the half-monostable circuit. IC2b triggers on the leading edge of the ring signal generated by IC2a. IC2c will trigger off the trailing edge of the same signal. To see this relationship more clearly, turn to Fig. 6-3. This illustration depicts that pin 2 of IC2b will generate a one-quarter second output pulse when triggered by the ground-to-positive transitions of the input signal (see Fig. 6-3B).

At the end of the standard telephone ringing signal, IC2c will detect and provide another one-quarter second pulse at its output but only at the trailing edge of the input ring signal. By combining these two pulses through diodes D2 and D3, you have the waveform illustrated in Fig. 6-3D.

At this point you have one pulse created at the beginning of the ring cycle and a second generated at the end of the same signal. By using these two pulses, you can control the gating (on/off) of an audio tone generator. As used in previous circuit designs, I have incorporated the use of another 4066 bilateral switch. If you can recall from the previous chapters, you can control the on and off period of a signal just by applying a positive or ground to the corresponding control pin. In the case of your electronic telephone ringer, this control pin is located at terminal 13, and the audio to be controlled is connected to pin 1. The audio output, like before, is connected to pin 2 of the 4066. The actual tone is created by XTAL1, a telephone receiver element.

THE TONE GENERATOR

With this arrangement, you can now control some sort of tone generator and synthetically create the English style, two-tone telephone signal. This tone can be created by the warble generator of the original English style telephone ringer (chapter 4) or the two 555 tone generators of chapter 5. But to create a telephone ringer project in a jiffy, use the audio generator illustrated in Fig. 6-4.

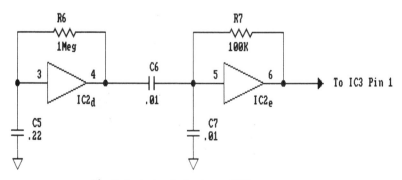

Fig. 6-4 *Instead of using a 555 to generate the needed tone for the ringer, use a 4584 hex inverter IC (Schmitt trigger).*

The generator does not reek of sophistication. But it does supply you with a pleasant two-tone warble audio output. And best of all, it can be built with the remaining unused sections of IC2 (IC2d and IC2e).

Just like the tone generator presented in chapter 5, this arrangement is divided into two distinctive elements. IC2d produces a train of pulses with a frequency of about 12 pps (pulses per second). IC2e generates an audio tone of about 500 Hz. When connected together by capacitor C6, the audio signal generated by IC2e can be literally turned on and off at a rate created by IC2d, thus producing a very simple warble audio output.

To complete your circuit, the audio output of IC2e, or any other tone generator can be connected to the bilateral switch (IC3 pin 1). The ring gating arrangement as previously discussed will control the overall on and off state of the device.

CONSTRUCTION

The parts needed to build the electronic telephone ringer are listed in the parts list. There are no exotic ICs or resistor values. If you are the owner of an electronic junk box, you might already have all the components needed to assemble your own telephone ringer. If not, appendix D will provide you with the address of a number of electronic mail order houses where these and other components can be purchased with the least amount of fuss. Send a postcard to a number of these businesses and request their latest catalog.

The schematic of the circuit in Fig. 6-2, reveals that the device is relatively easy and straightforward. For the experienced assembler, the construction of the ringer will take no time at all.

But even for the experienced technician, certain elements of construction must be met. The basic element is neatness. Do not mount components on a perf board as to let them dangle in mid air. Instead, mount resistors, capacitors, and diodes in such a way as to let them lie flat against the surface of the board.

Do not exaggerate the wire length between components, make all wiring connections short and direct. There is nothing more unsightly and troublesome than a bunch of brightly colored wires tangled up on the underside of a board, especially when it comes time to troubleshoot.

With these basic suggestions in mind, complete the construction of the electronic telephone ringer. Copy the template in Fig. 5-5 to use as a label.

Fig. 6-5 *Finish the electronic ringer with a fancy front-panel template.*

INSTALLATION OF THE ELECTRONIC TELEPHONE RINGER

As with all projects that are used as a substitute for the standard telephone bell, the ringer inside the phone must be somehow disconnected. If you own an imported telephone, all you have to do is to locate the small slide switch that is located just below the telephone keypad. Just flip this switch to its off position.

If you own an ITT or AT&T telephone, this procedure is a bit more complicated.

Not to bore you with the same long explanation on how to disconnect the telephone ringer from the line, just flip back a few pages to chapter 4, where a full explanation on disconnecting the bell is given.

Once the ringer inside your telephone is disabled, all that remains is to connect the electronic telephone ringer to the line. Take a small length of telephone line cord (this can be purchased at any local department store) and connect the red wire and connect it to the remaining side of capacitor C1. Connect pin 2 of IC1 to the green wire in the same cord.

Connect the other end of the cord to the telephone line. Here you have two options to consider. If your telephone installation is new, you will have the modern snap-in modular connectors on both your wall jack and on your telephone. To make use of this easy installation, the line cord that will be used to connect your project to the telephone line must be of the 1/4 modular type. Figure 1-31 of chapter 1 illustrates a cord of this type. If you wish to go this route, connect the wire leads from this modular cord as indicated above.

When soldered, you can take the modular end of the cord and snap it into an unused telephone jack. That is all there is to it. Small piano wires in the wall jack rub against metal contacts inside the line cord clip to create a complete circuit for normal current flow and a speedy installation for the installer.

The second option requires you to remove the cover of the telephone jack located on the wall using a screwdriver. Connect the red line cord wire to the red wall installation wire. Do the same with the green line cord wire. When complete, replace the jack cover. The telephone ringer is now connected in parallel to the line.

TESTING THE CIRCUIT

To test the operation of the ringer is quite easy. Just take a small screwdriver and short out pins 4 and 5 of IC1 (4N33 opto-isolator). When first shorted, you should hear a one-quarter second tone. Now remove the screwdriver. You will then hear another one-quarter second tone. What you have done is to simulate the incoming ring signal across the telephone line and to imitate the leading and trailing edges of an input pulse.

If all is working up to this point, all there remains to do is to have a friend call you on the telephone. When ringing, IC1 will sense the signal on the line and automatically turn on the tone ringer where you will be alerted to the fact that someone is determined to speak to you.

PARTS LIST FOR THE ELECTRONIC TELEPHONE RINGER

R1 R2	10 kΩ resistors
R3	2 kΩ resistors
R4 R5	2.7 KΩ resistors
R6	1 MΩ resistor
R7	100 kΩ
C1 C4	0.47 μF 250 V tubular capacitor
C2 C3	220 μF 35 V Electrolytic capacitor
C5	0.22 μF Disk capacitor
C6 C7	0.01 μF Disk capacitor
D1	1N4001 diode
D2 D3	1N914 diode
IC1	4N33 opto-isolator
IC2	4584 hex inverters (Schmitt trigger)
IC3	4066 bilateral switch
XTAL1	Crystal transducer (Telephone receiver element)
1	12 Vdc wall transformer
1	Telephone line cord
1	PC or perf board
1	Housing

The Melody Ringer II

TECHNOLOGY IS CHANGING RAPIDLY, BELLS USED IN TELEPHONES SEEM TO BE A THING of the 1950s. Specially designed integrated circuits have replaced the annoying mechanical device with the soft chirp of solid-state technology.

This magically transforms a standard bell telephone into a customized melody ringer. There is no need to purchase an imported telephone add-on device, because the Melody Ringer II can be constructed in about one-half hour, and the cash layout is so small that you will be surprised. Most of the parts needed might already be in your junk box.

There is only one component that you will not have in your junk box unless you are an avid electronics engineer. That missing part is the melody IC itself. Using an integrated circuit that can be easily purchased from Radio Shack, you can construct a device that plays one of 12 melodies or even one right after the other every time a telephone call is received. It all depends on the digital code that you install during construction. But more on that a little later.

HOW IT WORKS

For the following, refer to the schematic diagram presented in Fig. 7-1. With the melody ringer connected in parallel to the telephone line (red/green wires) as shown, ringing voltages in excess of 90 V ac (30 Hz) are rectified and brought down to a more appealing voltage by resistor R1. The voltage is delivered to pins 1 and 2 of the opto-isolator (4N33). As discussed previously on many other telephone projects, the 4N33 is a device that incorporates a small LED that flashes on a light-sensitive Darlington transistor circuit. With a voltage applied to pins 1 and 2 (LED input) by the bridge rectifier (D1 to D4), the output on pin 5 goes low.

With this in mind, let's examine what happens when a telephone call is received. A ringing signal is supplied by the local telephone company is applied to the circuit via the telephone line cord. It enters the melody ringer via a telephone line cord. During its travels, the ringing signal meets up with C1 and C2 (two 0.47 μF, 250 V capacitors). These capacitors are used to block the normal telephone dc voltage from coming into the IC (4N33), thus keeping its internal LED always on.

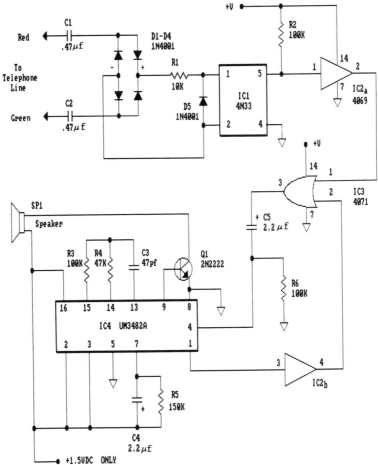

Fig. 7-1 *The schematic diagram for the melody ringer.*

Seeing that a standard LED draws a small amount of current, a 10 kΩ dropping resistor (R1) can be placed at the positive voltage output of a bridge rectifier (D1 to D4), where the ringing signal is converted into a pulsating dc voltage. With a 10 kΩ resistor in series with the telephone line (through the bridge rectifier circuit) the large pulsating dc is knocked down to acceptable voltage levels that can be handled easily by the LED inside IC1. This resistor can be 1/4 W carbon type.

If an oscilloscope is placed across pins 1 and 2 of the opto-isolator IC, you will see a pulsating dc voltage. This signal resembles a half-wave rectifier output. As indicated earlier, with a signal applied to pins 1 and 2 (IC1), the output (pin 5) goes low (logic 0). You can now use this ringing pulse to control the on status of the melody chip. But pin 4 of the chip requires a positive-going pulse to start everything off. So you must insert an inverter at this point. This is done by using one-sixth of a 4069 IC. Now with a negative-going pulse applied to the input of the inverter (IC2a) by pin 5 of IC1, the output of IC2a (pin 2) will create a positive-going signal.

If this pulse were applied directly to the melody IC, every time a ringing signal is detected across the telephone line (this happens every four seconds), the chip will automatically change the tune to the next in the sequence. After a while, this could become quite annoying because the IC will change tunes even though the first melody has not been completed.

To overcome the shortcomings of this type of design, you can make very good use of pin 1 of the melody chip. Pin 1 is called a *flag*. This means that the chip will signal the end of a melody by applying a positive voltage on this pin. You can make use of this flag by applying it to the input (pin 2) of a NAND Gate (IC3 4071). By applying the pulse generated by the incoming ring signal at pin 1 and the flag at pin 2 of IC3 (through a second inverter—IC2b), you can command the melody synthesizer to change tunes only at the completion of a current melody.

But what happens if the call is answered or if the caller hangs up the phone? This will create a situation in which there will be no additional ring pulses applied to the input of the NAND gate. With this scenario, pin 3 (output of IC3) will no longer apply a starting pulse to IC4, thus turning off the music generator at the end of the current melody.

Sounds interesting, doesn't it? But how does the melody chip work? What are the tunes it can play? What voltages are needed for operation? You probably have questions like these and many others. So set aside a page or two and examine closely the workings of this amazing integrated circuit.

THE UM3482A MELODY CHIP

The UM3482A is a multi-instrument melody generator using the newest mask-ROM (read-only memory) CMOS technology. The chip is designed to play melodies according to preprogrammed data. This data is inserted into the ROM memory during the manufacturing process. By using this technology, programmed data is permanently stored in the chip and does not require CPU (central processing unit—a computer) support to reload this information every time power is removed from the circuit.

With an internal memory capable of storing 512 seven-bit words, the UM3482A is capable of generating 12 songs (see Table 7-1) with three instrument sounds: the piano, the organ, and the mandolin (see Fig. 7-2).

Table 7-1. A listing of the tunes played by the UM3482A IC.

Selections:	
1) American Patrol	7) Are You Sleeping
2) Rabbits	8) Happy Birthday
3) Oh, My Darling Clementine	9) Joy Symphony
4) Butterfly	10) Home Sweet Home
5) London Bridges Falling Down	11) Wiegenlied
6) Row, Row, Row Your Boat	12) Melody On Purple Bamboo

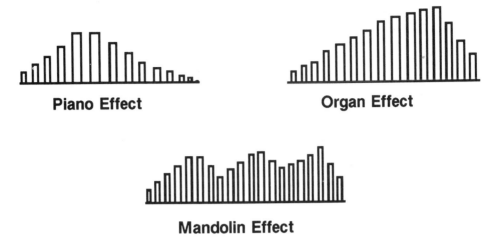

Piano Effect   **Organ Effect**

Mandolin Effect

Fig. 7-2 *By wiring the UM3482A IC in different ways, you can create a number of voices for the ringer.*

For a better understanding of the inner workings of this amazing IC, refer to the block diagram of the UM3482A in Fig. 7-3. Figure 7-4 illustrates the physical pin arrangement. To illustrate the function of each pin, Table 7-2 also is included.

In the introduction to this chapter, you read that the UM3482A allows you to program different modes of operation. To illustrate this, please refer to Table 7-3. By applying a ground at pin 2 (CE), the device is said to be in a standby mode. Note that the three Xs indicate what is called a "Don't care logic level."

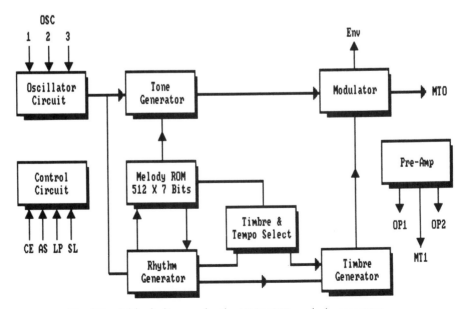

Fig. 7-3 *A block diagram for the UM3482A melody generator.*

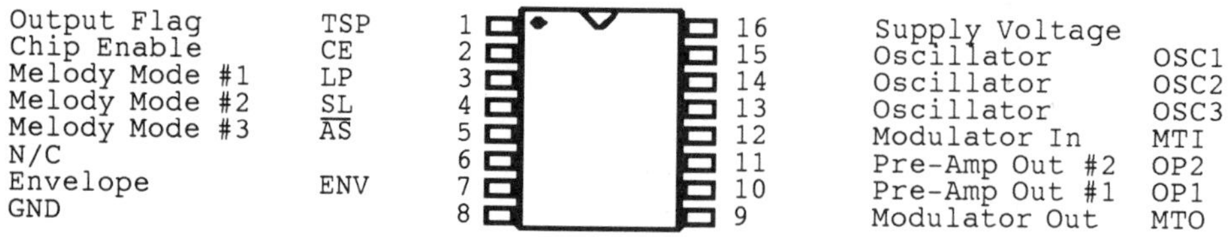

Output Flag	TSP	1	16 Supply Voltage
Chip Enable	CE	2	15 Oscillator OSC1
Melody Mode #1	LP	3	14 Oscillator OSC2
Melody Mode #2	SL	4	13 Oscillator OSC3
Melody Mode #3	$\overline{AS}$	5	12 Modulator In MTI
N/C		6	11 Pre-Amp Out #2 OP2
Envelope	ENV	7	10 Pre-Amp Out #1 OP1
GND		8	9 Modulator Out MTO

Fig. 7-4 *Pin assignments for the UM3482A melody generator IC (Radio Shack 276-1797).*

Table 7-2. A more in-depth listing of the pin assignments for the UM3482A music generator.

Pin Assignments and Functions

Pin	Assignments	Descriptions
1	TSP	Output flag of melody auto stop
		In normal operation this pin should be open
2	CE	Chip enable if connected to +1.5V
		Chip disabled if connected to GND
3	LP	Plays only one song if connected to +1.5V
		Plays all songs if connected to GND
4	SL	Positive going pulse to this pin will change melody to next song
5	$\overline{AS}$	Melody will repeat if connected to +1.5V
		Will auto stop if connected to GND
6	NC	No Connection
7	ENV	Envelope circuit terminal
8	GND	Ground return for power supply
9	MTO	Modulated tone signal output
10	OP1	Pre-Amplifier output #1
11	OP2	Pre-Amplifier output #2
12	MTI	Modulated signal inout to pre-amp
13	OSC3	External R-C oscillator
14	OSC2	External R-C oscillator
15	OSC1	External R-C oscillator
16	VDD	+1.5 Volt Supply Voltage

This means that the end function associated with item 1 (standby mode) can be created whether pins 3, 4, and 5 (LP, SL, and $\overline{AS}$) are at either a logic 1 or logic 0 level.

To continue the discussion, take a look at item 6 of the program truth table (Table 7-3). With a logic 1 at pin CE and a ground at pins LP and $\overline{AS}$, you can place a positive starting pulse at pin SL. The end results of this logic input will be that the UM3482A will first change to the next melody programmed into its memory and continue to play each song until the last is reached. Then, when the last melody is played, the IC will automatically stop.

By applying different logic levels on these programming pins, you can create different musical variations. If at this time, you turn your attention back to the schematic of the melody ringer II, you will notice that the UM3482A is programmed per item 8 of the program truth table (Table 7-3). This program will allow the IC to be in a standby mode until a positive-going pulse is sent to pin 4. As mentioned earlier, this pulse is generated by the incoming ring signal across the telephone line. When pin 4 receives this start pulse, IC4 (UM3482A) will

Table 7-3. To have control over the operation of the music generator, logic 1s, logic 0s and pulse signals must be placed at the correct pin locations.

Program Truth Table

Items	CE	SL	LP	$\overline{AS}$	Program
1	0	X	X	X	Stand-by
2	1	0	0	0	Start from first melody to last. Then stop.
3	⌐	0	0	1	Start from first melody to last. Then repeat from first.
4	⌐	0	1	0	Start from present melody. Then stop.
5	1	0	1	1	Repeat present melody.
6	1	⊓	0	0	Change to next melody. Play to last. Then stop.
7	1	⊓	0	1	Change to next melody. Play to last. Then repeat from first.
8	1	⊓	1	0	Change to next melody. Then stop.
9	1	⊓	1	1	Change to next melody. Then repeat same.

automatically change to the next programmed song. Then at the end of this tune, the IC will stop and wait for the next starting signal from IC3.

It is a good time to point out that a logic 1 to the melody generator is a + 1.5 Vdc. A logic 0 is, of course, a ground. Unlike other CMOS chips, the UM3482A can only be powered by a + 1.5 Vdc voltage. You might think that this requirement will complicate matters just a bit, but it is not. Figure 7-5 illustrates a suggested power supply that can be used by both the melody generator and the other CMOS ICs used in the Melody Ringer II circuitry.

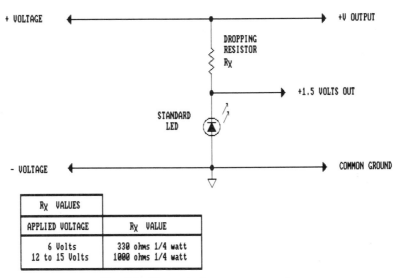

R_X VALUES	
APPLIED VOLTAGE	R_X VALUE
6 Volts	330 ohms 1/4 watt
12 to 15 Volts	1000 ohms 1/4 watt

Fig. 7-5   *The melody generator needs a + 1.5 V to operate. A problem you say? No, not by using this circuit.*

Say that you wish to power your melody ringer from a 6 Vdc supply. This voltage selection will prove to be adequate for all ICs concerned except for the UM3482A. Put a standard LED in the circuit with a dropping resistor (RX) in series. As an experiment, place a voltmeter across the LED leads. The meter will display a reading of about 1.5 V. What if two wires from IC4 were soldered across the LED leads? The voltage drop across the LED plus the low current required by the chip make this arrangement quite acceptable.

Note the resistor table also illustrated in Fig. 7-5. This table indicates the value of resistor RX for a variety of voltage inputs. Say you wish to power the circuit with a 6 V supply. From the table, the value of RX should be approximately 330 Ω with a wattage of 1/4 W. If the melody ringer is to be run at a higher voltage (12 to 15 V) (note that 15 Vdc is the maximum voltage that should be applied to a CMOS IC), the value of Rx will increase to 1000 Ω. Take note of these resistor variations and make the necessary corrections to the schematic illustrated in Fig. 7-5 before applying any voltage to the finished project.

In conclusion, to power the melody generator, just connect pins 5 and 8 to the common ground point. Apply the normal 6, 12, or 15 Vdc supply to all + V points. Finally, connect the junction of resistor RX and the LED to all required IC4 pins (see Fig. 7-1). With this arrangement, powering the melody ringer will be no problem at all.

Note that the suggested wiring of the melody generator as illustrated in Fig. 7-1 is a quick and dirty way to make the IC operable. Figures 7-6 and 7-7 illustrate two more complex wiring arrangements for the UM3482A. Whichever way you decide to wire the circuit, remember that the starting pulse from IC3 is always connected to pin 4. The busy flag (pin 1) is connected to pin 3 of IC2b.

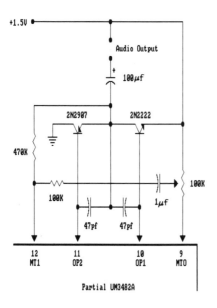

Fig. 7-6 *If you require a louder musical output from your melody generator, consider adding this two-transistor circuit.*

ASSEMBLY

Like the other projects in this book, the assembly of the Melody Ringer II is not critical. You can wire the ringer on a perf board or solderless experimenter board. Pay close attention to the polarity of C4, all diodes (D1 to D5), the LED (on the power supply), ICs, and of course the music synthesizer (IC4). If any

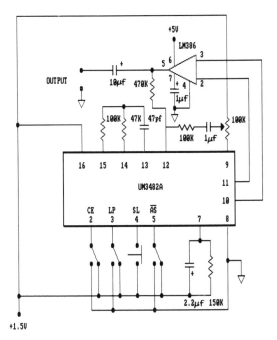

Fig. 7-7 *If you wish to manually control the operation of the music generator, you can make use of this switch arrangement. Also note that an LM386 amplifier has been connected to the audio output of the IC for added volume.*

NOTE: For added volume - connect a 10uf capacitor between
       pins 1 and 8 of the LM386 IC

problems develop in the operation of the melody ringer, look for improperly inserted components, solder bridges, and wiring errors.

If the problem still exists after a lengthy troubleshooting session, put the board aside for a while. Return to the ringer in an hour or two. You can be very surprised at the silly wiring errors you might have overlooked previously.

With the assembly of the ringer almost complete, remember that the UM3482A requires a + 1.5 Vdc for proper operation while the other ICs in the circuit can use 6 to 15 V. By applying a large voltage such as this to the melody generator, you can say bye-bye to the chip. So keep this critical requirement in mind in the final assembly of the circuit. Copy the template in Fig. 7-8 to use as a label.

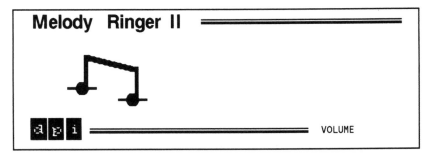

Fig. 7-8 *Front-panel template for the melody ringer.*

INSTALLATION AND TESTING

With the proper input voltage to the power supply, connect the melody ringer to the telephone line. This can be easily accomplished by using a TEE adapter. The TEE adapter is a labor-saving device that costs $2.00 to $3.00. For such a low price as this, it should be made part of your experimenter's junk box. For additional information on the TEE, just flip ahead to chapter 9.

With your Melody Ringer connected to the telephone line, short pins 4 and 5 of the opto-isolator (IC1) together. This short will provide you with a simulated ring signal. If all wiring is correct, you should hear, at this point, some sort of musical melody penetrating the air. At this time, remove the IC1 short. The melody will continue to play until the end is reached. Then it will automatically stop.

Re-apply the short between pins 4 and 5 of IC1, but this time remove the short as soon as the song begins to play. Now count to three. Then apply the short again. If the circuitry associated with the busy flag at pin 1 of IC4 is operating correctly, the melody will not change. The original song will play completely through without changing.

With this small but important test completed and the ringer determined to be operational, have a friend call you on the phone. With a ringing signal detected on the line by IC1, your Melody Ringer II will automatically kick into action and serenade you with one of 12 different melodies.

Remember that by using the Melody Ringer II, you now have no further need of the bell inside your telephone. So carefully remove the housing and trace out one of the four wires (if your telephone uses a 130BA ringer) or one of the two wires (if your telephone uses the 148BA ringer) from the bell. Once located, remove the wire from the screw or push-on terminal. Tape it and store it.

PARTS LIST FOR THE MELODY RINGER II

R1			10 kΩ resistor
R2	R3	R6	100 kΩ resistors
R4			47 kΩ resistor
R5			150 kΩ resistor

C1	C2	0.47 μF 250 V capacitors
C3		47 pF disk capacitor
C4	C5	2.2 μF 16 V electrolytic capacitor

Q1	2N2222 or PN2222 transistor

IC1	4N33 opto-isolator IC
IC2	4069 hex inverter IC
IC3	4071 OR gate
IC4	UM3482A music synthesizer (Radio Shack #276-1797)

D1 to D5	1N4001 or equivalent
SP1	8 Ω speaker
1	Telephone line cord
1	Special power supply—see text
1	Printed circuit or perf board
1	Housing

☎8

The Fone Sentry

IT IS 1:35 AM. THE COUNTING OF SHEEP EVENTUALLY PAID OFF. YOU HAVE FINALLY fallen asleep. The sense of reality has been lazily replaced with an imaginary dream world. Suddenly, your peaceful realm is shattered by the irritating clanging of a telephone. You hastily fling yourself at the ringing nuisance. You hesistantly reach for the receiver, even wondering; who will be calling at this hour? Within a second or so, the question asked by millions of people each and every night, is finally answered. A husky male voice at the other end of the telephone demands that you deliver a pizza with all the works and have it at his party within the next hour.

The preceding scenario was brought to you by the Fone Sentry. An easy to build, add-on telephone device that guards your telephone from all those late-night nuisance calls. By preventing your telephone from ringing until a prede-termined count is reached, unwanted calls will never again disturb your restful sleep. For those you wish to talk too, just inform them to let the phone ring from anywhere from 4 to 10 times. Once this predetermined count is reached, Fone Sentry will awake you with its built-in, pleasing tone ringer instead of the irritating clang of a bell.

Upon completion of the telephone conversation, Fone Sentry once again stands guard over your telephone line until the next nuisance call is received.

HOW THE FONE SENTRY OPERATES

The Fone Sentry works on the principle of a counting circuit. Instead of counting pulses in the neighborhood of millions per second, the Fone Sentry counts the number of times the telephone company places the ringing signal across the red and green wires of the line. When the count of the ring signals corresponds with that selected by you, Fone Sentry will enable a 4066 bilateral switch. The switch allows the internally generated ringer to penetrate the air with its pleasant-sounding, wabblelike tone signal.

Get involved with the inner workings of the Fone Sentry by referring to Fig. 8-1. This schematic illustrates that when a 120 Vdc wall transformer is con-nected to the indicated locations, IC1 (4069) will generate the same three tones produced by the English style telephone ringer (chapter 4). If you wish, you can flip back to chapter 4 to refresh your memory on the operation of the electronic tone generators.

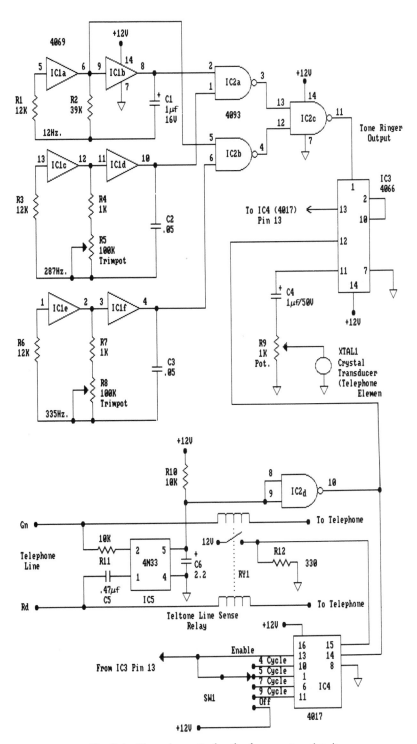

Fig. 8-1 *The schematic for the fone sentry circuit.*

As with the English telephone ringer, the secret of producing the pleasant sounding audio tones produced by the expensive NEC, Toshiba, and AT&T telephones, is the way the base frequencies are digitally mixed by the 4093 IC. The mixing of the base frequencies is the same as for the project presented in chapter 4.

To control the time that the fone sentry will signal an incoming call and to control the tone generator on and off sequence is determined by IC3, a 4066 bilateral switch. To see this unique switching arrangement more clearly, please refer to Fig. 8-2. Here, you can see that two of the four switches contained within the body of the 4066 IC are connected in series. From elementary electronic theory: for an output to appear at pin 11 of the IC, switch 1 and switch 2 must be closed. To close these switches, control voltages are introduced at pins 13 and 12 respectively. It is the job of these voltages to *mold* or *shape* the tone signal into the form that is needed.

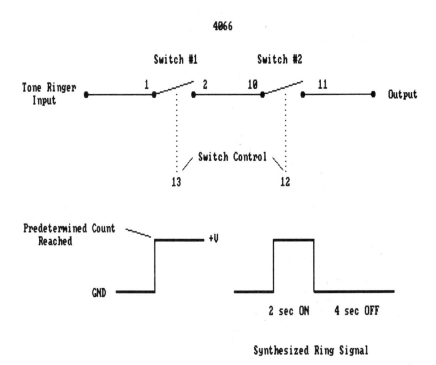

Fig. 8-2 *Controlling the 4066 bilateral switch (IC3) with positive-going pulses.*

To close switch 1, you must provide a positive voltage at pin 13 of the 4066. If you have an IC that counts the number of 110 V 30 Hz ring signals applied across our telephone line and have it go from ground to a positive voltage every time the predetermined ring count is reached. We can place this logic 1 at pin 13 to control the first half of the internally tone generator.

If switch 2 were not used, the audio tone produced by IC1 will be heard from XTAL1. But it would be a steady tone. A signalling tone of this type can quickly become very annoying. To compensate for this annoyance, switch 2 of the 4066 is used to shape the audio into the normal two-second on, four-second off ringing cycle (the industrial standard). If you take this incoming ring signal and condition it with IC5 and IC2d, you have the makings of the second needed gating pulse. Now, every time the ring signal across the telephone line is detected, these two ICs form a control pulse with an on period of two seconds and an off period of four seconds. It is this pulse that synthetically creates the telephone company's standard ring signal.

Getting back to Fig. 8-1, see just how these control pulses are created. For now, please turn your attention to the 4N33 IC (IC5). This opto-isolator is connected across the telephone line by using resistor R1 and capacitor C5. Internally connected across pins 1 and 2 of this IC is a small light-emitting diode. It is this LED that flashes when a ringing voltage is detected across the red and green telephone lines. When the LED is flashing, a light-dependent Darlington transistor array conducts, thus allowing the now-positive voltage applied to pin 5 from R10 to be grounded. To convert this pulse into a positive going signal, IC2d is used as an inverter. IC2d not only inverts the signal but the snap action associated with Schmitt triggers cleans up the pulse so it can also be used as a clocking signal for the 4017 (IC4). This brings you to the counting circuit.

The 4017 is a divide-by-10 counter with a 1-of-10 output. An interesting feature of this IC is that the 4017 contains an enable (pin 13) along with the standard counter reset (pin 15). With a ground placed at these two pins, the 4017 repeatingly clocks through its 10 output states. With each inputted clock pulse, the 4017 applies a positive voltage on its corresponding output. In other words, if the third clock pulse were to be entered, pin 7 (output 3) would have a positive voltage on it, and the remaining nine are grounded. When the fourth pulse is entered, pin 10 has the voltage, and pin 7, which once had the signal, will go to ground.

So think this out for a moment. If a ground is needed at pin 13 (and at pin 15) for the 4017 to count, what if you take a positive signal that is created by the counting circuit and apply it to the enable pin? If you do that, you can have the counter automatically stop at a predetermined count. Sound familiar? You bet.

So if you take pins 1, 6, 10, and 11 and connect them to a rotary switch (SW1) as shown in Fig. 8-1, you can apply the positive pulse from any counter output to the enable pin 13. You can, in effect, stop the 4017 from counting once the predetermined cycle is met. You can also take this positive pulse and connect it to the control pin 13 of the 4066 IC. This would close the first half of the series connected switches (switch 1 and switch 2—Fig. 8-2).

Now with the first half of the 4066 receiving a control voltage, all that is left to do is to get an audio output is to apply a second gating voltage to pin 12 of IC3. This is done by stealing the output from pin 10 of IC2d. With this voltage applied, the wabble tone generated by the Fone Sentry is heard from the crystal transducer (XTAL1).

When the telephone is answered, you need some way of resetting the internal counters of the 4017 IC. If you remember, that pin 15 of IC4 must be at ground in order for the counting procedure to take place. If a positive voltage were placed at this pin, counting will immediately cease, and counters would be cleared back to zero (0). At this point, the next clock pulse received will be counted as pulse 1.

You can create this reset voltage very easily. By using a special telecommunications device called a line-sense relay (see Fig. 8-3) made by Teltone Corporation, you can sense the lifting of the telephone headset and have a contact inside the relay close. It is the closing of this relay that applies the needed reset voltage to IC4 pin 15. When the handset is returned to the cradle, the contact of the line-sense relay opens up, disconnecting the + 12 Vdc from pin 15 and reapplying the ground through the 330 Ω resistor (R12). At this point, your Fone Sentry is rearmed and ready to prevent those annoying calls.

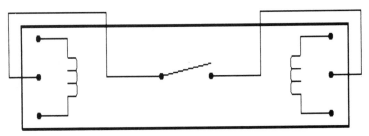

Fig. 8-3 *An inside look at the Teltones line-sense relay.*

CONSTRUCTION

By using the computer-generated printed-circuit board layout, you can easily assemble the Fone Sentry with the investment of one evening of time.

The circuit is straightforward and does not require the purchasing of any exotic ICs or components (except for the Teltone line-sense relay). If you prefer, you can assemble the circuit on a perf board. No special precautions must be considered, just that if point-to-point wiring is used, remember to keep the 12 V power supply leads away from the telephone lines. The phone company will not like the idea of you adding an additional 12 V signal to their 48 V line, so be careful.

For your convenience, the Teltone line-sense relay and the etched and drilled printed circuit board can be purchased from Del-Phone Industries, Spring Hill, Florida, just as many of the specialized ICs used in this book can be. Del-Phone Industries can provide the needed material to assemble a working project. Their address is given in the parts list. Just drop them a self-addressed, stamped envelope for the pricing on these and many other items of interest.

Whichever you decide on: purchasing or making your own PC board from the computer-generated artwork presented in Fig. 8-4, note that the board requires that you install six jumpers. These connections must be made before installing any components. As you can see from the parts layout (Fig. 8-5), one

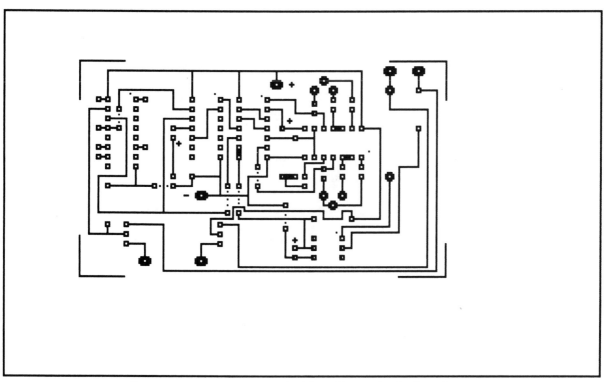

Fig. 8-4   *A computer-generated PC board layout for the Fone Sentry.*

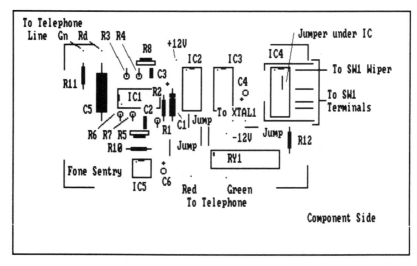

Fig. 8-5   *The parts placement diagram for the sentry PC board.*

of the jumpers is located under IC4. This location makes it imperative that you install the jumper before installing the 4017.

The resistors that make up the Fone Sentry should be the next items to be installed and soldered. The capacitors are the third item. Just note that capacitors C1, C4, and C6 are electrolytic, and they do require that you install them the correct way. To help stuff the capacitors to their proper orientation, note the small plus (+) sign etched on the board. If you line up the positive side of each capacitor with that symbol before soldering, you will have no problem at all.

The final components to be installed are the five ICs. Just like the electrolytic capacitors, ICs are polarity-sensitive devices that can be placed in a circuit in only one way: the correct way.

Again, to help you with the proper orientation of the ICs, note the small dots just under selected IC holes. Each dot represents that this pad is for the number 1 IC pin. So for proper orientation, just keep in mind that the dot is pin 1.

The relay is the last component to be mounted on the board. Carefully insert the six legs into their holes and solder in place. Potentiometer R9 is the sentry volume control, and it should be a front-panel mounting with some sort of knob to allow for easy adjustments. The transducer (XTAL1), which reproduces the audio output, is an high-impedance crystal device. If you wish, you can substitute a telephone receiver element for XTAL1. The output produced by the telephone receiver will produce plenty of volume. Just remember to drill a number of holes in the housing to allow the audio to penetrate into the air.

To assist you in the wiring of the rotary switch, Fig. 8-6 is included. This illustration depicts the wiring of SW1. Note that the 4, 5, 7, and 9 output pins from IC4 have been pre-selected to allow you a greater degree of flexibility in determining the ring delay of the Fone Sentry. The fifth position of SW1 provides a steady positive voltage to the enable pin of IC4. This position will defeat the guarding capability of the Fone Sentry and will allow the telephone to ring on the first signal detected.

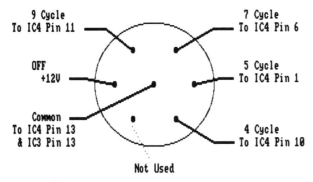

Fig. 8-6 *Wiring the SW1 rotary switch (rear view).*

THE POWER SUPPLY

The power required for the proper operation of the Fone Sentry is provided by a 12 Vdc wall transformer. A fully regulated power supply is not needed and installing one would only increase the overall cost of the project. Just remember to purchase a UL-approved device rated at 12 Vdc with about 200 mA of current.

When ready to install the power leads, just remember that the banded wire lead is the positive side and the installation into its correct location is critical for the proper operation of the fone sentry.

INSTALLATION OF THE FONE SENTRY

The fone sentry is a device that is installed in series with your existent telephone hookup (see Fig. 8-7). This means that you must disconnect the guarded telephone from the wall and reconnect it to the PC board at the two locations designated "To Telephone." Wiring polarity must be enforced throughout the installation. So when connecting the Fone Sentry to the line, remember to connect all red wires to red and green to all green. This would assure the continued operation of older tone-dial telephones. It is these phones that do not have built-in polarity guards like their newer counterparts.

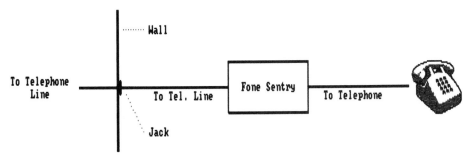

Fig. 8-7 *The Fone Sentry is a device that is placed in series with the telephone line and the telephone to be secured.*

Seeing that the fone guard has a built-in tone ringer, you must disable the ringer inside your guarded telephone. If the phone is an import model, all you must do is to switch the ringer off. This can be accomplished by flipping the slide switch located just under the keypad to off.

If you have a bell-type ringer, you must remove the housing and locate at least one of the wires from the ringer and remove it from the terminal where it is now residing. With the wire removed, you must use electrical tape on the lug to prevent any shorts inside the phone. When taped, just replace the housing and you are in business. Copy the template in Fig. 8-8 to use as a label.

TESTING THE FONE SENTRY

With the installation of the Fone Sentry complete, the time has come to test out the circuit. To test the tone ringer-circuitry, temporally connect a telephone

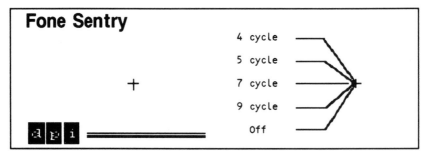

Fig. 8-8 *Front-panel template for the Fone Sentry.*

receiver element (through a 1 μF capacitor) to pin 11 of IC2. At this point, you should be able to hear some sort of sound. Using a frequency counter, place its hot lead at pin 10 of IC1 (while its black lead is at ground). Adjust R5 until a reading of 287 Hz is obtained. Now place the counter hot lead on IC1 pin 4 and adjust R8 until you see a frequency of 335 Hz. When these adjustments are made, you should hear a pleasant wabble output. With this test complete, remove the telephone receiver from IC2 pin 11.

For the next test, you need the help of a friend. With SW1 in its 4-cycle position, have your friend call you on the phone. With a scope connected to pin 10 of IC2, note the detection of the incoming ring signal. At this time, you should not hear any sound from the sentry. Count off the pulses. On the fourth count, the Fone Sentry will come alive with the soothing audio output of the tone ringer.

Answer the call and note a reset voltage on pin 15 of IC4. Have your friend call again, this time flip SW1 to its 9-cycle position. Again, count the number of detected incoming ring signals. On the ninth pulse, Fone Sentry will come alive again.

The last test is to check to see if the Fone Sentry can be over-ridden. This can be tested by placing SW1 in its off position. Have a friend call again. This time Fone Sentry will ring when it detects the first incoming ring pulse.

This completes the testing of the Fone Sentry. It is now time to sleep peacefully without worrying about all those after-midnight telephone calls. Just remember to tell your friends that the Fone Sentry is on guard at your house.

PARTS LIST FOR THE FONE SENTRY

R1	R3	R6	12 kΩ resistor
R2			39 kΩ resistor
R4	R7		1 kΩ resistor
R5	R8		100 kΩ trim pot (Digi-Key #KOA15)
R9			1 kΩ potentiometer (audio taper)
R10	R11		10 kΩ resistors
R12			330 kΩ resistor

C1	1 μF 16 V axial capacitor
C2 C3	0.05 μF disk capacitor
C4	1 μF 50 V radial capacitor
C5	0.47 μF 250 V axial capacitor
C6	2.2 μF 16 V radial capacitor
IC1	4069 hex inverter IC
IC2	4093 quad 2-input NAND Schmitt trigger IC
IC3	4066 bilateral switch IC
IC4	4017 divide-by-10 counter IC
IC5	4N33 opto-isolator IC
XTAL1	Crystal transducer (or telephone receiver element)
SW1	1-pole 6-position rotary switch (5)
RY1	Teltone line-sense relay*
1	12 Vdc wall transformer
1	Telephone (tone-dial or rotary)
1	Telephone line cord
1	Printed circuit board (or perf board)
1	Housing

---

*Available from Del-Phone Industries, P.O. Box 5835, Spring Hill, Florida 34606

The Hi-Tech MOH (Music-on-Hold) Adapter*

THE THOUGHT OF PLACING A TELEPHONE CALL ON HOLD, IS BY NO MEANS, A FRESH, innovative concept. Large institutions have been using this idea for a number of years. Only until the court-ordered breakup of ITT, little black boxes have been designed by foreign manufacturers and presented to the eagerly awaiting public. Using names like MOH (music-on-hold), tele-hold, and telephone hold button, the general public was offered the same telephone conveniences that large corporations have been blessed with for a number of years.

By combining another modern miracle, the miniature music synthesizer, with a basic telephone hold button, you can serenade your caller with an unexpected musical interlude until reconnected by you. Placing a call on hold or injecting an audio signal into a telephone line is a lot easier than you might think.

Present in this chapter is a circuit called the hi-tech music-on-hold adapter. No, this adapter is not a reprint of an SCR (silicon-controlled rectifier) circuit; it is a new concept that allows you to place calls on hold and impress a musical signal on the line. By using the standard DTMF signals of a conventional tone-dial telephone (2500 or 3554 home telephone) as the controlling element for the internal electronics of this project, you can very easily place a call on hold and then reconnect it. All this can be accomplished just by pressing the * (press this button to place a call on hold) and the # (press this button to reconnect the holding party) buttons located on a tone dial. But of course, before you can succeed in such an endeavor, you have to be familiar with the basics of DTMF transmission and decoding as well as how to simulate a resistive load across a line without even using a telephone instrument.

---

*Reprinted with permission of *Radio Electronics* (July 1989)

THEORY OF PLACING A CALL ON HOLD

The normal on-hook voltage across a conventional telephone line at about 50 V. This voltage might be different for you. It all depends on a number of factors. This includes your location within the country as well as the local telephone company's central office equipment. For the sake of simplicity, assume this voltage is 50 V. When you lift the handset from its cradle, the impedance of the telephone (about 600 Ω) is placed across the line, causing the voltage to drop to 5 V. For an illustration of this drop, please refer to Fig. 9-1. The secret of placing calls on hold is to fool the telephone company's central office switching equipment into thinking that a telephone is still in use, when in reality it is not. To do this, simply connect a resistive load across the line using a controllable relay. In the case of the hi-tech music-on-hold adapter, this resistive load is in the form of a 120 Ω, 1 W resistor.

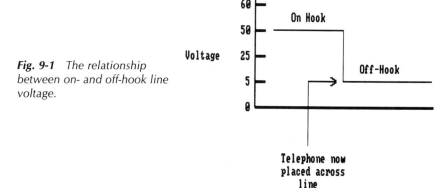

Fig. 9-1   *The relationship between on- and off-hook line voltage.*

In addition to placing a resistor across the line, some sort of on-hold indicator must be used in order to inform you that there is a call on hold. For such an easy task, a standard LED is pressed into service. To light the LED, no external power is required (although a voltage of 5 V is needed to power the ICs of the adapter). Inspection of the schematic in Fig. 9-2 reveals that the LED is wired in series with the 120 Ω, 1 W resistor and the contacts of your control relay. When connected across the telephone line by the closing action of the control relay, telephone line voltage (5 V) is now flowing through the resistor, LED, and relay contact combination. With the correct polarity applied from the telephone line (tip/ring), the LED will light, indicating that a call has been placed on hold. The lighting of the LED will also allow current to flow through the resistor, thus completing the electrical path and fooling the central office into believing that a phone is off-hook. Also, take note that there is a 1.5 V drop across the LED. If you wish to include the optional music synthesizer, use this voltage to power the module (Z1). So with this scheme, the lighting of the LED and the powering of the synthesizer is being supplied by the kind people of the telephone company.

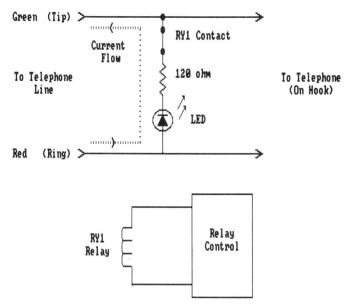

Fig. 9-2 *By connecting a resistive load across
a telephone line, the on line party can be placed on hold.*

DUAL TONE MULTIFREQUENCY

Tone-dial telephones such as the conventional 2500, 3554, and 2554 (which can be found in almost any home), produce a special kind of signal called DTMF, which stands for dual tone multifrequency. The concept of DTMF dialing has already been discussed and it is recommended that you flip back to chapter 1 to refresh your memory regarding this new telecommunications media.

THE DTMF RECEIVER IC

The G8870 receiver IC is an extremely complicated device. Older DTMF receivers required large and bulky audio filters. The G8870 incorporates all the needed filtering within its 18-pin body. Unlike its counterpart, the Teltone M-957-01 IC, used in the digital telephone lock project, the G8870 does not need additional circuitry that would allow it to be connected across the telephone line. The M-957 requires the addition of an op-amp and other components. The 8870 includes an internal op-amp. To simplify construction of the hi-tech music-on-hold adapter and to keep the cost of construction to a bare minimum, use the 8870 (a block diagram of the 8870 can be found in Fig. 9-3) instead of the M-957. Extensive tests of the adapter circuit have found no appreciable difference in use, operation, or circuit reliability by using the 8870 over the M-957.

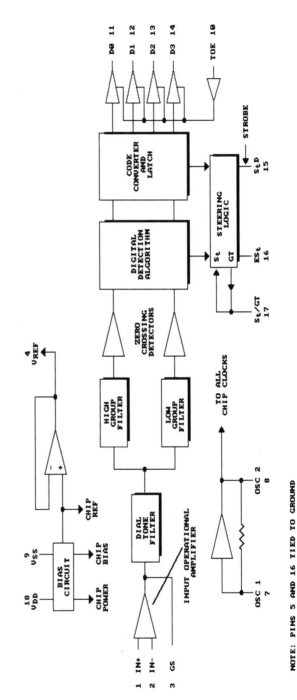

Fig. 9-3   *A block diagram of the G8870 DTMF receiver IC.*

HOW IT WORKS

The hi-tech MOH adapter is a complex circuit from the point of view of an electronic layperson (see Fig. 9-4). For the advanced hobbyist, the adapter also presents a challenge even though I have included the printed board artwork in Fig. 9-5.

Before you start any assembly, you need to know how the circuit operates and the additional circuitry that can be included to give your finished adapter that store-bought look.

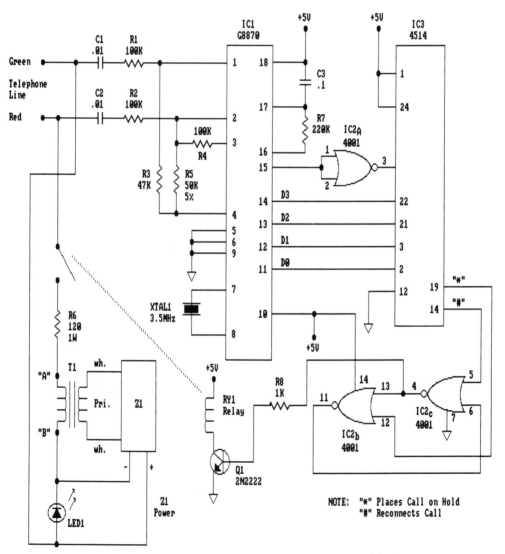

Fig. 9-4 *The schematic diagram of the high-tech music on-hold adapter.*
(Reprinted with permission from Radio Electronics Magazine, *August 1989 issue.*
© Copyright Gernsback Publications, Inc.)

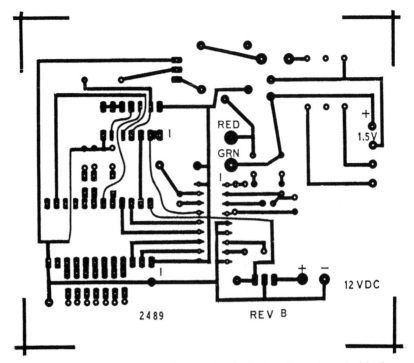

Fig. 9-5   *Artwork to the scale of 1 to 1 for the high-tech music-on-hold adapter.*

First of all, the adapter is connected in parallel across the green and red wires of a standard telephone line where it sits quietly until you activate its circuitry with a press of the ∗ button (obviously, you would need a tone-dial type telephone for this circuit to operate properly).

By pressing this button, two tones are generated by the tone dial of the telephone. They are 941 Hz and 1209 Hz. As discussed in chapter 1, these tones are combined to form the dual tone multifrequency signal. It is this signal that is picked off the telephone line by C1, R1, C2, and R2, where it is applied to the internal operational amplifier of the DTMF Receiver integrated circuit (G8870/IC1). At this point, the dual 941 and 1209 Hz tone is converted into its binary equivalent (1110 binary). This code can be seen at pins 11, 12, 13, and 14. Pin 11 is D0 while pin 14 is the D3 data output. Pin 15 of the G8870 is the strobe output (or StD on the block diagram). This pin will present a logic high when a received tone pair (a valid DTMF signal) has been registered and the output latch of the IC has been updated. The StD pin (or strobe) will return to a logic low in the absence of an input signal.

At this point, you have decoded the pressing of the tone dials ∗ button into its binary equivalent (1100) and you also have a logic high (logic 1) at pin 15 (strobe or StD output). But before this data can be made into a useful output, you must further decode this raw electrical signal. To do this needed conversion, IC3 is used. This 1-of-16 decoder (4514) produces a logic high at the pin that corresponds with the inputted binary equivalent. In this case, when the ∗

button is pressed on the telephone, a binary code of 1100 is applied to the 1, 2, 4, and 8 input terminals of the 4514 (pins 2, 3, 21, 22 respectively). When the inhibit pin is brought low (0), pin 19 of IC3 is made to go high (1). To further explain this operation, assume that you have pressed the # button of the tone dial. Now the binary output of the DTMF Receiver is 1110. This data is then converted by IC3 into its numerical value of 12 (or # of a tone dial). This can be easily seen by connecting a VOM to pin 14 of the 4514. This conversion also applies to the 10 other possible tone-dial combinations that can be applied. Although the 4514 can decode any binary data up to and including the number 15 (pin 15).

As mentioned, pin 23 (inhibit) of IC3 must be made low so that the 4514 can decode the pressing of the * button on the telephone. Seeing that there is a logic high (1) at the strobe output of IC1 (pin 15) and you need a logic low at pin 23 of IC3, a simple inverter can be used at this point.

Not to make the circuit overly complicated by using an entire inverter IC package for this one operation, you can construct this digital building block by tying together two unused inputs of a NOR gate. This gate can be found in IC2 (4001 quad 2-input NOR gate). When connected per the schematic, you now have the needed low (0) to apply to pin 23 of IC3.

THE NOR LATCH

The heart of the MOH adapter is the triggered NOR Latch. This latch is made up of two independent NOR gates wired as seen in Fig. 9-4. When a logic high (1) is applied to pin 5, the circuit latches in a state such that there is an output voltage at pin 4. But if a high (1) were applied to pin 12 of IC2, the latch will reset, thus forcing pin 4 to produce a low (0) output. If you connect a relay (RY1) to the output of this all important Latch through a driving transistor (Q1), you can control the relay on and off states with the use of the tone dial of a telephone. All this can be accomplished by pressing the * button to activate RY1 while pressing the # button will release RY1. By controlling the relay in this fashion, you can easily wire the relay to connect or disconnect the 120 Ω resistor (R6) from the telephone line, thus holding or reconnecting the party on the line at will.

POWER SUPPLY

The template for the MOH adapter also includes an on-board 5 V power supply. This simple circuit (see Fig. 9-6) is powered by a UL-approved dc wall transformer. The voltage of the transformer can be anything from 6 to 15 Vdc at about 200 mA current rating. The needed 5 Vdc is derived from the 7805 voltage regulator IC (IC4). This IC not only knocks down the applied voltage to the needed 5 V but also regulates the output up to 1 A. Capacitor C4 (470 μF, 16 V) is an electrolytic capacitor used to filter any ac component that may still be present in the power line.

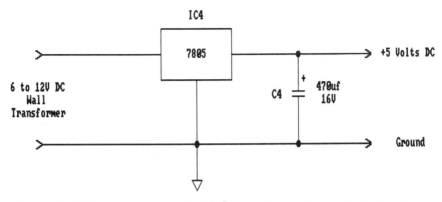

Fig. 9-6 *Can't forget a power supply. Here is one that can be used with the adapter.*

CONSTRUCTION

The hi-tech music-on-hold adapter is assembled on a single-sided PC board; the layout for the board is provided in Fig. 9-7. Alternately, an etched and drilled board can be purchased from the source given in the parts list. Using Fig. 9-7 as a reference, install the resistors and capacitors first, then the solid-state devices and the relay. If you wish, use sockets or strip-sockets for all ICs, (but they are really not needed). Just quickly solder them into place. If done correctly, there is very little chance that you will damage the heat-sensitive integrated circuits. Of course, all four ICs, all capacitors, transistor Q1, and the LED are polarity-sensitive electronic devices. Use care while installing them. Double check their orientation against the parts layout diagram (Fig. 9-7) before soldering.

The completed unit can be installed in just about any kind of cabinet. The size of cabinet depends on whether or not you decide to include the circuit options (options are discussed a little later).

Fig. 9-7 *If you make the PC board from the provided artwork, you will need to know where to put the parts.*

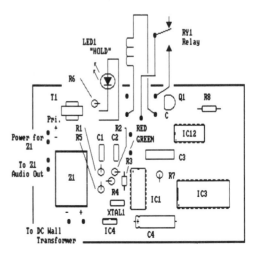

It is recommended that the listed wall transformer be a UL-approved type and rated at a current rating not to exceed 1 A. Using this type of UL device, you will prevent any shock hazard to the user or yourself and also prevent, in the case of a serious mis-wire, any ac voltage from entering the line. An ac voltage could cause serious problems for the telephone company.

MOH OPTIONS

To make this circuit truly a MOH adapter, you must include some sort of melody generator or connect the adapter directly to an AM/FM receiver. The latter might seem to be more involved than needed, particularly when foreign manufacturers have developed micro-miniature music synthesizers that play a repeating 30 second tune with only the addition of a 1.5 V battery and an audio transducer.

These truly remarkable circuits can be easily modified to operate with our adapter. Figure 9-8 illustrates the simple changes that must be made to the synthesizer. First, start by removing the battery and make note of its polarity while in the circuit. The battery is no longer needed, so it can be discarded or saved for another project you might have in mind.

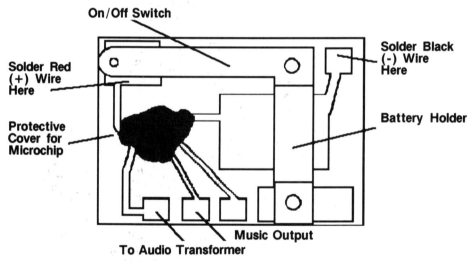

Fig. 9-8 *The additional wiring needed to provide power and audio output from the music synthesizer module.*

Taking a short piece of red wire, solder this lead to the positive (+) side of the battery holder. Using a short piece of black wire, solder this to the negative (–) side of the battery holder. Finally, cut off the audio transducer at the transducer side of the two white wires. Last, your synthesizer contains a rather simple on/off switch. This switch must be soldered closed. If not, oxidation will form on the contacts over time, thus allowing intermittent operation. To prevent this, just use a small amount of solder to permanently short out the contacts.

Interconnection of the music module to the printed-circuit board is quite simple. Note in Fig. 9-7 there is extra board space for the mounting of the synthesizer. Mounting in this reserved space should be accomplished by using double-sided tape about 1/4-inch wide. Further investigation of Fig. 9-7 will reveal the designation 1.5 V etched in the copper along with a (+) sign. At this point, solder the red wire from the synthesizer. Solder the black wire to the next lower hole. The final two holes are used to interconnect the audio output from the module to the primary of the Mouser TM013 transformer.

The secondary of transformer T1 is connected, via the copper traces, in series with the holding resistor (R6) and the LED (LED1). With an audio signal being impressed on the secondary from the music synthesizer, it is combined with the normal 5 Vdc off-hook telephone line voltage. This combination allows the party on hold to be serenaded by the synthesizer rather than listening to the inanimate stillness of a dead telephone line.

If the use of the music synthesizer is not an important feature to you, a small length of bus wire must be soldered in place of the secondary winding of transformer T1. This wire will allow the current flow from resistor R6 to continue its journey to the LED. Otherwise an open circuit will exist.

The LED (LED1) is recommended and should be part of your finished project. Whether you make use of the musical synthesizer or not, this LED will give you a visual indication that a call has been placed on hold. So its inclusion is a Must.

Finally, to give your finished adapter that professional appearance, use the panel template in Fig. 9-9. This template can be copied from the page, cut out, then glued onto the front panel of your housing.

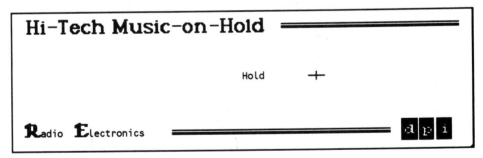

Fig. 9-9 *Template for the finished adapter.*

INSTALLATION

The installation of your finished adapter is a very simple matter. Using a telephone line cord that can be purchased at any local Radio Shack store, connect the red and green line cord wires to the appropriate location of the MOH PC board. The yellow and black leads are not used and should be taped and stored. Determine the location of the telephone you wish to have access to the MOH adapter. When you have decided, using a screwdriver, open up the cover of the telephone jack that is mounted on the baseboard. Parallel connect the red wire from the adapter to the red telephone line. Do the same for the green adapter wire. The yellow and the black wires found in the telephone jack are not used. They should be left alone. Connect the other end of the line cord to the telephone junction box.

An easier way to install your adapter is to use a special connector device, called a TEE adapter (see Fig. 9-10). This adapter is used on newer telephone installations that make use of the modular line cords. This adapter allows you simply to snap two telephone modular line cords, (in this case, one telephone and one MOH adapter) into one telephone jack. This scheme prevents you from opening up the telephone jack and perhaps introducing a short across the line.

To use the modular line cord installation concept, just visit your local Radio Shack store and purchase a line cord with a modular connector on one end and space lugs on the other. This cord is called a 1/4 modular line cord.

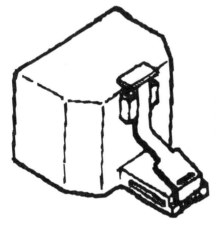

Fig. 9-10 *To connect two devices to a single telephone line, you can purchase a TEE adapter.*

As discussed above, connect the spade lugs (red/green) to their appropriate location on the adapter PC board (see Fig. 9-11). With the other end, just remove the telephone modular line cord and reconnect this clip into one of the TEE adapter available openings. Then snap the adapter cord into the remaining opening. Finally, take the TEE adapter and reconnect to the telephone wall jack.

That is all there is to it.

TESTING AND OPERATION

To test the finished MOH adapter, just dial a friend as you would normally. When your friend is on line, press the * button on the tone dial. Relay RY1 should pull in, and if you made use of the musical optional, a tune also should be heard in the handset. You can now hang up the telephone receiver without disconnecting your call. The LED (LED1) should also be on.

To reconnect a call on Hold, lift up the receiver then just press the # button on the tone dial. Relay RY1 will now drop out, thus disconnecting resistor R6 from the line. At this time, the LED will go out and the music will cease.

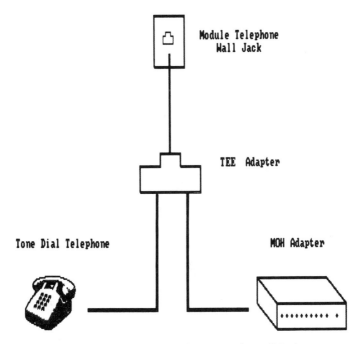

Fig. 9-11 *By connecting the TEE to the wall jack,*
a telephone and your finished MOH adapter can be connected to the same line.

Your hi-tech music-on-hold adapter is now sitting patiently waiting for your next command.

PARTS LIST FOR THE HI-TECH MUSIC-ON-HOLD ADAPTER

R1 R2 R4	100 kΩ resistors
R3	47 kΩ resistor
R5	50 kΩ 5 percent resistor
R6	120 Ω, 1 W resistor
R7	220 kΩ resistor
R8	1000 Ω resistor
C1 C2	0.01 μF disk capacitor
C3	0.1 μF disk capacitor
C4	470 μF 16 V electrolytic capacitor, axial type
T1	Audio transformer, 1 kΩ primary—8 Ω secondary (Mouser Elec. 42TM013)
T2	6 to 15 V 250 mAdc UL wall transformer
IC1	G8870 DTMF receiver IC*
IC2	4001 NOR gate

IC3	4514 1-of-16 decoder (high output)
IC4	7805 5 V voltage regulator TO-220
LED1	Standard red LED
RY1	5 or 6 V SPST (single pole, single throw) or SPDT (single pole, double throw) relay (100 Ω coil)
Q1	2N2222 transistor
XTAL1	3.58 MHz crystal*
Z1	Music synthesizer module*
1	Telephone line cord (spade tipped or modular type)
1	TEE adapter optional*
1	PC board*
	Wire, solder
1	Housing

---

*Available from Del-Phone Industries, P.O. Box 5835, Spring Hill, Florida 34606.

A Flip/Flop Hold Button

CHAPTER 9 OF THIS BOOK PRESENTS A MORE ADVANCED HOLDING CIRCUIT USING A specifically designed integrated circuit, the G8870 DTMF receiver. This chapter introduces you to a circuit that is somewhere between the sophistication of a DTMF circuit and the simplicity of an SCR circuit.

HOW IT WORKS

This version of the telephone hold button makes use of an IC called a dual D flip/flop (4013). Although this IC contains two independent flip/flop circuits, the F/F (flip/flop) hold button will make use of only one. Figure 10-1 illustrates quite well, the pinout of this CMOS integrated circuit.

The dual D flip/flop is configured as a counting circuit. As an example, apply a positive going pulse to the clocking input (pin 3 or pin 11) of the 4013. The internal circuitry will count this pulse and apply a logic 1 to the Q output (pin 1 or pin 13), and the $\overline{Q}$ output (pin 2 or pin 12) will be grounded.

If this positive pulse were applied to the base of an NPN switching transistor, the device would turn on, allowing the relay connected to its collector leg to pull in.

Apply another positive pulse to the clocking pin of the IC. This time the internal circuitry will detect this pulse and flip the circuit back into its resting position. This position, of course, will be a state where the Q output will be low (logic 0) and the $\overline{Q}$ output will be high (logical 1).

Again, if this Q output were being delivered to the base of a switching transistor, the transistor will now be in its off state, causing the relay connected to its collector leg to drop out.

In simpler terms, you have a circuit in which a relay pulls in when you press a button the first time. When you press the same button again, the relay drops out. With a circuit of this type, you can literally control the placing of a holding resistor across the telephone line.

For a clearer understanding, look at Fig. 10-2. Push button (SW1) is the relay control switch. But the schematic shows that by pressing SW1, a grounding signal is generated at the junction of resistor R1 and capacitor C1. This is the correct logic needed by this circuit because you must refine the pressing of the button in two ways. First, you must have one and only one ground pulse generated by the pressing of SW1. Without this circuit, SW1 will generate a

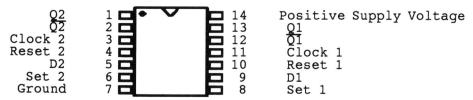

Q2	1		14	Positive Supply Voltage
Q̄2	2		13	Q1
Clock 2	3		12	Q̄1
Reset 2	4		11	Clock 1
D2	5		10	Reset 1
Set 2	6		9	D1
Ground	7		8	Set 1

Fig. 10-1 *Pinout assignments for the 4013 dual D flip/flop.*

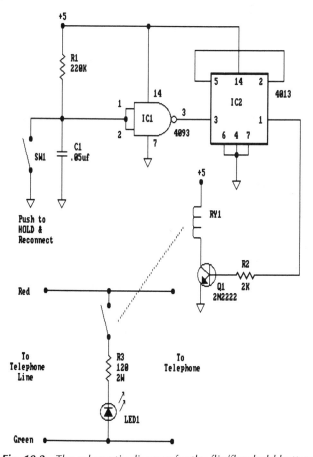

Fig. 10-2 *The schematic diagram for the flip/flop hold button.*

series of make and break pulse. This is caused by *contact bounce.* Contact bounce is an intrinsic problem with all types of switches. To correct this inborn switch problem, you can have a capacitor of sufficient size to filter out all but one output pulse. This filtering arrangement is the job of capacitor C1. This 0.05 μF capacitor charges up to its peak voltage through the resistance of R1 (220 kΩ). When a short is applied across the cap by pressing SW1, it discharges in about 100 milliseconds. This capacitor discharge masks the associated contact bounce of SW1, allowing only one positive to ground signal to be generated.

The second refinement you need is a circuit that will give a fast, snap-action on signal as well as a fast off. You also need to invert the created output pulse. Both requirements can be met by making use of a Schmitt trigger. To be exact, the 4093 (2-input NAND Schmitt trigger).

IC 4093 will provide a snap-action pulse when its input voltage exceeds 2.9 V (5.9 V if a supply voltage of 10 V were used). The IC provides snap-action off pulse when the input voltage falls below 2.3 V (3.9 V if a supply voltage of 10 V were used). Thus the *hysteresis* or noise immunity of this dead area is 0.6 V with a 5 V supply and 2 V with a 10 V supply.

The inversion of the created pulse is also a simple task. You can short two input pins of the NAND gate together. With this arrangement, a NAND gate can be made to simulate an inverter.

The holding circuit is a collection of inexpensive components. By using a 120 Ω, 2 W resistor, you can fool the telephone company central office equipment into believing that a telephone is connected across the line even when it is not. This is the basis of a hold button. The rest of the circuit consists of the relay control and a visual indicator.

The job of the indicator is to signal the user that a call is placed on hold. In reference to the schematic illustrated in Fig. 10-2, this indicator is in the form of a standard red LED. As you can see from the schematic, this LED is connected in series with the RY1 relay contact and the 120 Ω resistor. When the relay is activated by pressing the SW1 button, the series circuit composed of these components is placed in parallel across the telephone line (of course, when this relay is closed, the user can hang up the phone without worrying about disconnecting the call).

With this series circuit now across the line, the dc voltage being applied to the line by the telephone company is flowing through the components. This voltage will allow the LED to light up, thus indicating that the RY1 relay has pulled in and that a call is on hold.

To reconnect the party, just press SW1 a second time. This will de-energize the relay, thus removing the 120 Ω resistor from the line. Just remember to lift the telephone handset from the cradle before pressing SW1.

MUSIC OPTIONS

The schematic presented in Fig. 10-2, illustrates a basic telephone holding circuit. With an arrangement such as this, once the RY1 relay pulls in, the party on hold will be greeted with the lifeless sound of silence. Why subject your caller to the stillness of an inactive telephone line? An alternative is the tranquilizing effect of a music synthesizer. Chapter 9 shows you how to modify a micro synthesizer into a project usable circuit, and chapter 7 describes how to use the UM3482A music generator. With minor modifications to the existing F/F hold button circuit, you can easily include the soothing effects of a musical interlude to your project.

Figure 10-3 depicts the changes needed to create a sophisticated holding circuit. By adding an easily obtainable audio transformer (1 kΩ primary—8 Ω

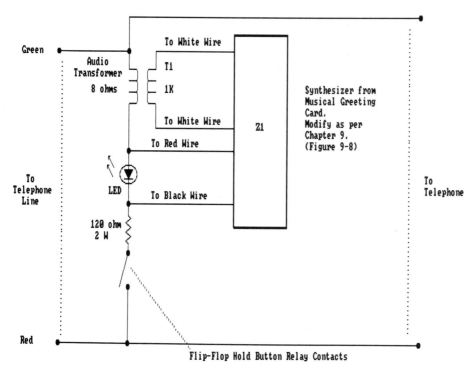

Fig. 10-3 *If you wish to add a musical melody
while a call is on hold, just wire the F/F hold button as shown here.*

secondary) and a micro synthesizer, purchased in any card shop as part of a
musical greeting card, you can easily impress an audio signal across any telephone line.

Referring to Fig. 10-3, note that the 120 Ω resistor, LED and the RY1 relay
contacts are still in series. All that was added in this circuit was the 8 Ω secondary winding of an audio-matching transformer. The 1 kΩ primary winding of
this transformer is then connected to the two white wires of the greeting card
synthesizer. The audio transformer in this circuit is used as an isolation device.
It prevents direct connection of the synthesizer to the telephone line.

The synthesizer itself must have a voltage of 1.5 V to operate. Just as
explained in chapter 9, this module can be powered by the telephone line by
connecting its positive and negative voltage leads across the series circuit LED.
For a more in-depth explanation of the modifications needed by the micro
music synthesizer, please refer to Fig. 9-8.

Figure 10-4 illustrates another music generator that can be easily incorporated into the F/F hold button. Just as discussed in chapter 7, the UM3482A can
also be called into service. As with the micro music synthesizer, the only modification needed to the existing hold circuit is the addition of the miniature
audio transformer. Note that the transformer used in Fig. 10-3 is also used here,
but it is reversed. Now the audio output of the UM3482A is connected to the
8 Ω secondary winding, and the 1 kΩ primary is included in the series-holding
circuit.

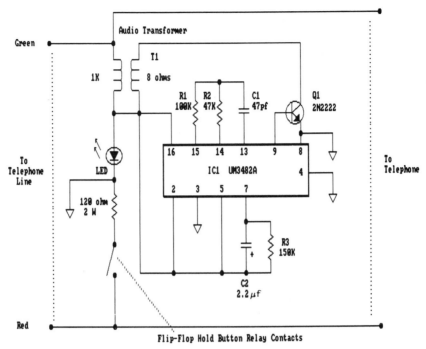

Fig. 10-4 *You can also add the UM3482A music generator to the basic hold circuit.*

A closer inspection of the UM3482A shows that the chip programming has changed. As shown in Fig. 10-4, pins 2 and 5 are connected to a logic 1 (+ 1.5 V), and pin 3 is grounded. By referring to Table 7-3, you can see that a programming input of this nature will allow the IC to change to the next melody and play to the last. Then it will allow the IC to repeat, playing from the first melody.

Table 7-3 also shows that pin 4 (SL) should be a positive-going pulse, but in the schematic SL is grounded. To refresh your memory, refer to Table 7-2. The SL pin assignment for the UM3482A IC is a melody-changing pulse. Seeing that you have already programmed your melody generator to change tunes automatically, a pulse on pin 4 is not needed, just a ground is needed.

As with the greeting card melody generator, the UM3482A requires a + 1.5 V for its operation. So to power the generator, we can use the same LED wiring arrangement as with the micro synthesizer. That is, to steal the 1.5 V needed for operation from the telephone company.

So as the circuit stands, the UM3482A will effortlessly play one melody right after the other and impress its output across the telephone line through the action of the audio transformer.

CONSTRUCTION

As with many other projects in this book, the F/F hold button circuit has no critical parts or assemblies to worry about. Because there are three options available to the assembler, printed circuit board artwork was not included. After you have decided on which options are to be incorporated into your final design, a PC board might not be a bad idea. It will prevent wiring errors, prolong the life expectancy of the circuit, and provide a professional appearance.

If etching a printed circuit board is not your strong point, the next best route is to use the perforated board. These boards are extremely useful in the construction and troubleshooting of your project. These boards can be purchased in just about any electronics mail order house as well as any Radio Shack store.

Whether you make your own PCB or use the perforated board, the construction is straightforward. There are no exotic ICs or other components to purchase. Probably all the components are already on hand in your electronic junk box. So before going out and buying parts, the junk bin is the best place to start looking for that needed component.

With all parts on hand, begin to assemble the hold button. The two ICs should be mounted on sockets. These sockets will give you a very good starting place to mount the resistors and capacitors. The only polarity-sensitive components that you should be aware of are the two ICs, LED1, and transistor Q1. These parts can be wired only one way. Otherwise you will be defeated even before you begin. The circuit will just refuse to operate. So pay strict attention when mounting these components.

If you wish to include one of the two musical options as illustrated in Figs. 10-3 and 10-4, the complexity of the circuit will also increase. With an increase of complexity also comes the increased possibility of wiring mistakes. So be careful. Take your time. Time taken now will be time saved in the troubleshooting of the finished project. Copy the template in Fig. 10-5 to use as a label.

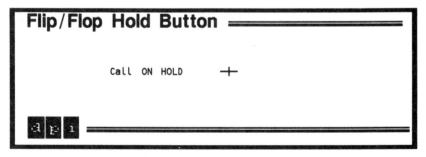

Fig. 10-5 *With your F/F hold button complete, dress up the project with this front panel template.*

POWERING THE CIRCUIT

The flip/flop hold button uses CMOS technology. These ICs can be powered by any voltage from 3 to 15 Vdc. Figure 10-2 shows a recommendation of using a 5 V power supply. If you wish, you can increase this voltage up to and including 15 V. But be aware that if you increase this operating voltage, you must increase the voltage rating of the RY1 relay coil. Presently, the relay is a 5 to 6 Vdc coil with a SPDT (single pole, single throw) contact. To prevent any problems associated an increase in voltage levels, stay with the recommended value.

The recommended voltage needed for operation can be obtained from a UL-approved wall transformer with an output of 5 Vdc (100 mA). Voltage regulation is not necessary and should not be considered. It will just increase the overall cost of the project.

TESTING AND INSTALLATION

To test your finished F/F hold button, it need not be connected to the telephone line. Just apply a 5 Vdc level to all points indicated on the schematic. With voltage applied, press the spring-return push button (SW1). When it is pressed, you should hear as well as see relay RY1 pull in. Press the button again, the relay should now drop out.

If you have added one of the two musical options, test the generators next. To do this, connect the negative voltage lead from a power supply delivering 5 V to the "To Telephone Line" red connection on the board. Then connect the positive voltage lead to the telephone line green connection. When connected, power up the circuit and press SW1.

With SW1 pressed, three things happen. First, the relay will pull in. Second, the LED will light. Finally, if you connect an oscilloscope to the audio transformer, you will see the tune being generated by the UM3482A IC or the greeting card music synthesizer module. If any problems develop, immediately remove power from the circuit. It is now time to troubleshoot the circuit.

To troubleshoot this or any other circuit you must obviously look for mis-wires. Shorted IC pins caused by wire leads that are too long and solder bridges (or splashes) should also be considered. The last thing to be considered is that the IC is defective. ICs purchased from reliable sources are of first quality and they should always be considered bad as a last resort. If your troubleshooting points to a defective IC, it must be replaced. For this reason, all first projects must be constructed with IC sockets. If no sockets were used, you will have to desolder and remove all wires from each pin of the IC. This is not only time consuming but even if the IC were good to start out with, the constant heat being created by desoldering the device will, without a doubt, destroy the delicate component. If sockets were used to start out with, all you have to do is to remove power, pry up the IC, exchange it with a replacement component and re-apply power. If the project starts to work, the problem has been troubleshot down to a defective integrated circuit.

After needed troubleshooting, your hold button project finally works. If it took a little time to troubleshoot, that is OK. Just think of what you have learned. Remember it for another day.

So connect this thing to the telephone line and give it a whirl. Using a telephone line cord, connect the hold button to the wall jack, making sure you observe the polarity of the wires. Apply power. The time has come. Call a friend on the phone. With a conversation in progress, press SW1. You should hear the relay kicking in. Also you should hear a musical interlude penetrating through the telephone headset. The LED should be on. At this time, you can hang up the telephone.

To reconnect the call, just lift the headset from the cradle. Then press SW1 again. The relay will pop up. The music will stop. The LED indicator will go out. At this time, you can now carry on with a normal conversation. If you wish to place the call back on hold again, just press SW1 anytime you wish.

PARTS LIST FOR THE FLIP/FLOP HOLD BUTTON

R1	220 kΩ resistor
R2	2 kΩ resistor
R3	120 Ω 2 W carbon resistor
C1	0.05 μF disk capacitor
Q1	2N2222 transistor
IC1	4093 Schmitt trigger NAND gate
IC2	Dual D flip/flop
SW1	Spring-return push-button switch
LED1	Standard red LED
RY1	5 or 6 Vdc SPST relay
1	Printed circuit or perforated board
1	Telephone line cord (modular or spade tip)
1	Housing
1	dc wall transformer (5 Vdc 100 mA)

PARTS LIST FOR MUSIC-ON-HOLD OPTION (MUSICAL GREETING CARD MODULE)*

T1	Audio transformer 8 Ω – 1 kΩ (Mouser Electronics)
Z1	Music synthesizer from musical greeting card modified per Fig. 9-8
1	Wire

*For proper operation, use these parts in addition to the components listed in the flip/flop hold button parts list.

PARTS LIST FOR MUSIC-ON-HOLD OPTION
(UM3482A MUSIC GENERATOR INTEGRATED CIRCUIT)*

R1	100 kΩ resistor
R2	47 kΩ resistor
R3	150 kΩ resistor
C1	47 pF disk capacitor
C2	2.2 μF 6 Vdc electrolytic capacitor
IC1	UM3482A music generator (Radio Shack)
Q1	2N2222 transistor
T1	Audio transformer 8 Ω – 1 kΩ (Mouser Electronics)

---

*For proper operation, use these parts in addition to the components listed in the flip/flop hold button parts list.

☎11
Dialed-Digit Display (LED Light-Emitting Diode Output)

THE NEXT THREE PROJECTS PRESENTED ARE DIALING DISPLAYS. THESE DISPLAYS ARE circuits that allow you to verify, in some way, a telephone number as it is being dialed. This confirmation is to be made before the switching equipment of the local telephone company has a chance to kick in. This will, of course, prevent the billing of telephone numbers.

The first of the three projects presented allows you to verify, by the use of LEDs (light-emitting diodes), the number as it is being dialed on a tone telephone. To help in the assembly of the device, a computer-generated layout for the complete dialed digit display is included.

HOW IT WORKS

Figure 11-1 shows the complete display device. A close inspection reveals that the circuit uses the G8870 DTMF IC. This chip, as you will recall from other chapters, allows you to connect your circuit directly across the telephone line without worrying about balancing, phone line interfacing, and equipment isolation. Just by adding a few relatively inexpensive resistors and capacitors you can accomplish the same function available with the Teltone M-957-01 IC.

As mentioned in chapter 9 (hi-tech MOH), the G8870 accepts the incoming tones or frequencies generated by the telephone dial pad and speedily converts them into their individual binary equivalents (available at pins 11, 12, 13, and 14). Pin 15 of the DTMF IC delivers a *strobe*. This strobe line produces a logic 1 when a valid tonal input is received.

You can make use of this strobing effect by enabling pin 23 of a 4514 (1-of-16 decoder with high output). Then, with a binary code at its input and its strobe pin low, the 4514 can provide, once decoded, a logic 1 at the output of the associated pin. To further illustrate the operation of the 4514, Fig. 11-2 shows the pinout of the IC, and Table 11-1 shows the truth table for various binary inputs.

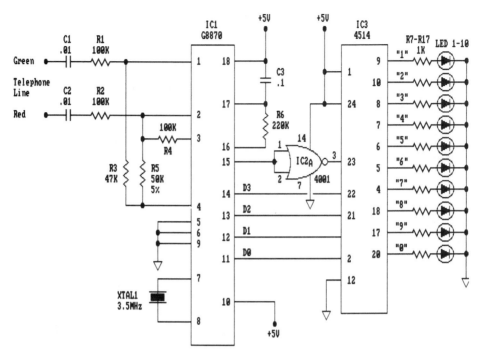

Fig. 11-1   *Schematic diagram for the dialed-digit display using an LED output.*

Follow	1		24	Supply Voltage
Data Input D0	2		23	Inhibit
Data Input D1	3		22	Data Input D2
Output #7	4		21	Data Input D3
Output #6	5		20	Output #10
Output #5	6		19	Output #11
Output #4	7		18	Output #8
Output #3	8		17	Output #9
Output #1	9		16	Output #14
Output #2	10		15	Output #15
Output #0	11		14	Output #12
Ground	12		13	Output #13

Fig. 11-2   *Pinout assignments for the 1-of-16 decoder IC (4514).*

Because binary numbers can be decoded into their corresponding decimal equivalents by using the 4514, small lights or in this case LEDs can be made to magically light on command. With the proper legend on the faceplate of a plastic cabinet, you can easily see the telephone number as it is being dialed.

THE LED (LIGHT-EMITTING DIODE)

A number of projects make use of the LED, so turn your attention for a brief moment to a description of it.

Table 11-1. The 4514 truth table.

	Input						1-of-16 High Output															
Pin	22	21	3	2	1	23	11	9	10	8	7	6	5	4	18	17	20	19	14	13	16	15
Function	D3	D2	D1	D0	Follow	Inhibit	Decoded Output															
Decimal																						
1	0	0	0	0	1	0	1	0	0	0	0	0	0	0	0	0	0	0	0	0	0	0
2	0	0	0	1	1	0	0	1	0	0	0	0	0	0	0	0	0	0	0	0	0	0
3	0	0	1	0	1	0	0	0	1	0	0	0	0	0	0	0	0	0	0	0	0	0
4	0	0	1	1	1	0	0	0	0	1	0	0	0	0	0	0	0	0	0	0	0	0
5	0	1	0	0	1	0	0	0	0	0	1	0	0	0	0	0	0	0	0	0	0	0
6	0	1	0	1	1	0	0	0	0	0	0	1	0	0	0	0	0	0	0	0	0	0
7	0	1	1	0	1	0	0	0	0	0	0	0	1	0	0	0	0	0	0	0	0	0
8	0	1	1	1	1	0	0	0	0	0	0	0	0	1	0	0	0	0	0	0	0	0
9	1	0	0	0	1	0	0	0	0	0	0	0	0	0	1	0	0	0	0	0	0	0
10	1	0	0	1	1	0	0	0	0	0	0	0	0	0	0	1	0	0	0	0	0	0
11	1	0	1	0	1	0	0	0	0	0	0	0	0	0	0	0	1	0	0	0	0	0
12	1	0	1	1	1	0	0	0	0	0	0	0	0	0	0	0	0	1	0	0	0	0
13	1	1	0	0	1	0	0	0	0	0	0	0	0	0	0	0	0	0	1	0	0	0
14	1	1	0	1	1	0	0	0	0	0	0	0	0	0	0	0	0	0	0	1	0	0
15	1	1	1	0	1	0	0	0	0	0	0	0	0	0	0	0	0	0	0	0	1	0
16	1	1	1	1	1	0	0	0	0	0	0	0	0	0	0	0	0	0	0	0	0	1

How an LED Works

Just as a common glass diode is made up of a piece of P and N type semiconductor material sandwiched together, so is the LED. When forward biased with an electrical current, electrons are forced to travel from one material to the other. This transition results in the formation of heat as well as light. Because the light created by this electron flow is only as small as a pin head, a means of light amplification had to be created before any useful application could be considered.

The most economical means of light amplification is a common glass lens (see Fig. 11-3). By mounting the PN *semiconductor* material at the bottom of a tubular can, a glass lens can be supported just above the light-producing material. Now, when the device is forward biased, the pin point of light is magnified by the lens, thus illuminating the entire top of the LED.

Just like diodes and transistors, LEDs are polarity-sensitive devices. This means that LEDs require the proper voltage polarities on their leads. Figure 11-4 shows the internal structure of a standard red LED. Note that the two leads protruding from the main body have the names *anode* and *cathode*. Also note the length of the cathode lead with respect to the anode leads: the cathode lead is much shorter.

The second way to determine which lead is which is to locate the flat side of the *casing*. This flat side represents the cathode lead. To forward bias a device of this type, the cathode lead is to be connected to ground, and the anode is to be connected to the positive side of the available power supply. Just

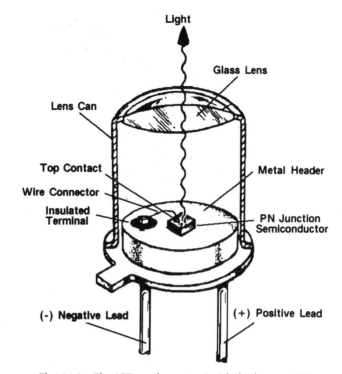

Fig. 11-3 *The LED package. An inside look at an LED.*

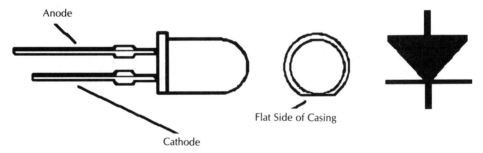

Fig. 11-4 *Just like a diode, an LED also has polarity.*

remember that an LED produces a relatively small resistance level when forward biased, so the addition of a *dropping resistor* is a must. By referring back to Fig. 11-1, you can see that the dropping resistors are made up of R7 to R17. These 1 kΩ resistors prevent an abnormal high current from entering the LED and destroying the delicate PN junction.

By allowing the 4514 to produce a logic 1 at one of its output pins, you can take this voltage and deliver it to an LED through a resistor. It is this changing output voltage level that lights the appropriate LED when the telephone is being dialed.

ASSEMBLY

Assembly of the dialed-digit display is not extremely critical. It can be constructed on a perf board, or you can make use of the computer-generated artwork presented in Fig. 11-5. Using the provided artwork will, without a doubt, increase the overall cost of the project, but you will be rewarded with a finished project that will be both pleasing to the eye and a more rugged assembly. If you prefer to fabricate your PC board or send the enclosed artwork to a vendor, you will need the component layout diagram illustrated in Fig. 11-6. Note that a jumper wire must be soldered just to the right of resistor R6. This jumper provides a positive 5 V to pin 24 of IC3. This connection must be made for proper operation.

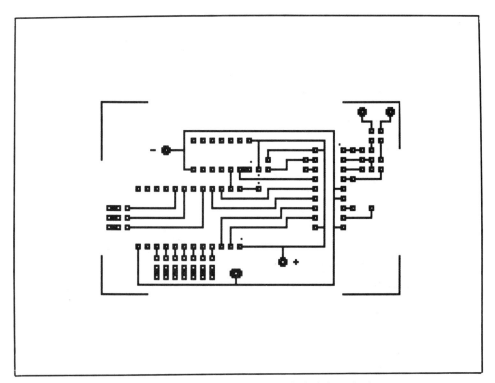

Fig. 11-5 *A computer-generated layout for the dialed-digit display (LED output).*

The only polarity-sensitive components used on the digit display are the integrated circuits and the ten LEDs. As for electrolytic capacitors, do not waste your time looking—none are needed. C1, C2, and C3 are all inexpensive disk capacitors with a voltage rating of about 12 V. The values of these components are standard, over-the-counter capacitance values. They pose no problem in purchasing.

All resistors used in this circuit are carbon composition types with a tolerance of 20 percent. The only difficulty that you might encounter is with R5.

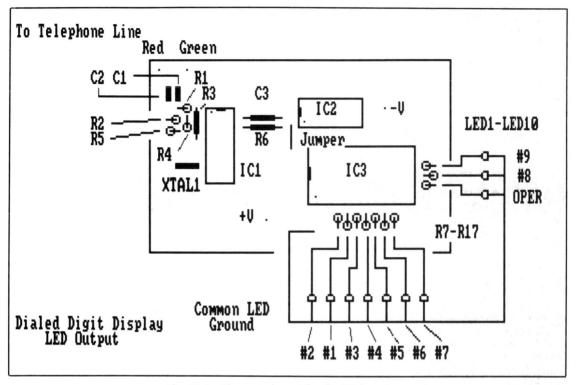

Fig. 11-6   *The parts layout for the PC board.*

This resistor is a 50 kΩ, ¹/₄ W carbon resistor, but the tolerance must be within 5 percent. So if you are the type of technician that likes to substitute resistor values in project designs, you may tamper with R1 to R4, but leave R5 alone. It must remain 50 kΩ, ¹/₄ W, 5 percent.

As shown on the parts placement diagram illustrated in Fig. 11-6, all components are mounted on the circuit board except for the 10 LEDs. These devices are mounted on the front panel of the chosen housing. The cabinet can be made of either metal or plastic—the choice is up to you. Using a plastic housing that can be found in a Radio Shack store is the ideal medium for drilling. Drilling a hole in soft plastic is much easier as well as safer than drilling in metal.

With the construction almost complete, spruce up the appearance of the final assembly by making use of the front panel template illustrated in Fig. 11-7. Copy this template on a copier, trim it to fit the panel, and glue it in place. Templates of this type can save you an hour or more if you wish to include front-panel text as well as graphic illustrations.

TESTING AND INSTALLATION

With the dialed-digit display assembled, check for solder bridges and shorts. Also inspect the assembly for mis-wires (if you make use of a perf board), incorrect component values, and improperly installed polarity sensitive components (ICs and LEDs).

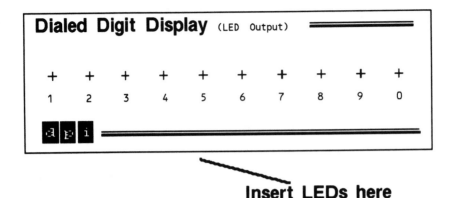

Fig. 11-7   *Template for the front panel label.*

After you have inspected the board and you are satisfied that there are no potentially dangerous shorts on the board, apply the operating voltage at the indicated points. With a voltmeter, check the following IC pins for a 5 Vdc reading:

IC1	IC2	IC3
Pin 18	Pin 14	Pin 1
10		24

check for grounds at the following locations:

IC1	IC2	IC3
Pin 5	Pin 7	Pin 12
6		
9		

and at all LED cathodes.

With all the correct voltages applied and all grounds connected, temporarily disconnect power from the digit display. With a 6- or 7-foot length of $1/4$ modular telephone line cord, solder the green wire to the free end of capacitor C1. Then solder the red wire from the line cord to the free end of capacitor C2. The other end of the modular cord is then easily snapped into the new modular wall jack (a TEE adapter might be needed at this point). In reality what you have done is to connect the digit display in parallel with the telephone line.

With the circuit now installed across the telephone, the time has come to put the thing through its paces. With the power supply delivering the needed 5 V, lift the handset of the telephone and press each of the 10 tone-dial buttons. As you do, the corresponding LED will light. That is all there is to it.

With the now-operating dialed-digit display mounted on its attractive housing, you can proudly show off your handicraft as it sits next to your telephone waiting to serve your dialing needs.

PARTS LIST FOR THE DIALED-DIGIT DISPLAY (LED OUTPUT)

R1 R2 R4	100 kΩ resistors
R3	47 kΩ resistor
R5	50 kΩ resistor 5 percent carbon
R6	220 kΩ resistor
R7 to R17	1 kΩ resistors
C1 C2	0.01 μF disk capacitor
C3	0.1 μF disk capacitor
LED1 to	
LED10	Standard red LED
IC1	G8870 DTMF receiver*
IC2	4001 NOR gate
IC3	4514 1-of-16 decoder (high-output)
1	$^1/_4$ modular telephone line cord
1	Housing
1	5 V dc power supply

---

*The G8870 IC is available from Del-Phone Industries, P.O. Box 5835, Spring Hill, Florida 34606.

☎12
Dialed Digit Display (Seven-Segment Light-Emitting Diode) Output

CHAPTER 11 SHOWS HOW YOU CAN TAKE AN ORDINARY LED AND HAVE IT ACT AS AN indicator that visually tells you the telephone number that is being dialed. You can take this concept one step further. Build a circuit that converts the binary output of a DTMF receiver chip (G8870) and have it drive a seven-segment LED readout.

This is an all-important chapter. Read it to understand not only the action of converting a binary code into a more pleasing and useful display but also to grasp the theory of programming *memory chips*. These electronic brains are used more extensively in upcoming projects like the talking telephone.

Figure 12-1 illustrates the new dialed-digit display. Closer scrutiny of the schematic shows a number of parts that have not been used before. This includes the seven-segment LED display, the memory chip (IC2), and the binary decoder/driver IC (IC3). By combining these easily obtainable parts, you can construct a unique telephone dialing aid.

Before you dive into the theory and construction of the dialed-digit display, examine the individual operations of these newest electronic components to be used in your customized telephone circuits.

THE SEVEN-SEGMENT READOUT

Just as common LEDs are biased to produce visible light, so are seven-segment readouts (or displays). A seven-segment readout, in reality, contains seven individual LEDs formed into bars by the manufacturer. These bars are then laid end to end to create a figure 8.

LED readouts also have a common connection point, and you are required to apply either a ground or logic 1. A voltage of the opposite polarity is applied to the individual segments to light them. By lighting discreet segments, you can create any number from 0 to 9 (and some letters). To standardize the production of LEDs, manufacturers have assigned letters to indicate each of the seven

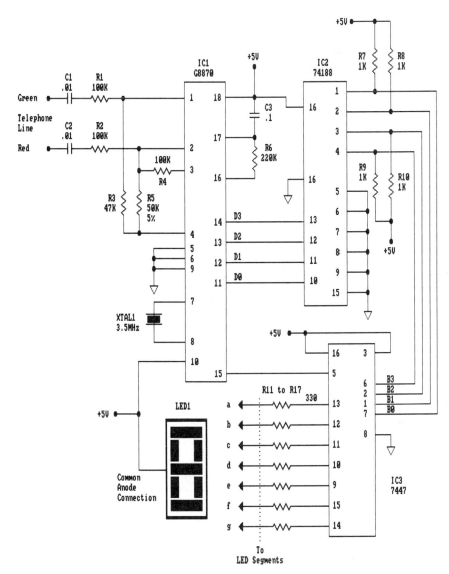

Fig. 12-1   *The schematic diagram for the dialed-digit display using a seven-segment LED output.*

segments. Figure 12-2 shows the typical shape of a readout and the letters associated with each segment. By flipping between Fig. 12-2 and Table 12-1, you will begin to see a pattern emerging from all this. By applying the correct polarity between segments, you can develop a means of allowing quick interpretation of binary data.

Figure 12-3 illustrates two commonly available LED readout types. Figure 12-3A shows a common-cathode readout. To light the individual segments on this readout, the common connection point must be connected to ground (or

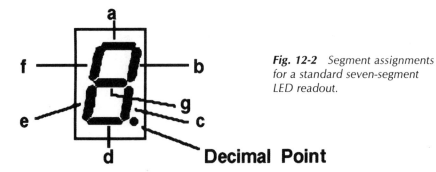

Fig. 12-2 *Segment assignments for a standard seven-segment LED readout.*

Table 12-1. To produce a display on the readout, individual segments are lit. This table shows the relationship between segments and the display it produces.

Character	Segment Assignments	Display
0	a,b,c,d,e,f	0
1	a,b	1
2	a,b,d,e,g	2
3	a,b,c,d,g	3
4	b,c,f,g	4
5	a,c,d,f,g	5
6	c,d,e,f,g	6
7	a,b,c	7
8	a,b,c,d,e,f,g	8
9	a,b,c,f,g	9

logic 0), and the desired segment termination point (letters a to g) must have the opposite polarity applied. In this case, you must deliver a +5 V.

Conversely, Fig. 12-3B requires that the common connection point be wired to the positive side of the voltage source, and the individual segments are connected to ground (or logic 0). In the case of Fig. 12-3A, this device is called a common-cathode seven-segment readout or display. Figure 12-3B is called a *common-anode* seven-segment readout or display. Both of these can be purchased at Radio Shack, Digi-key, or any other mail order electronic store.

Seeing that your digit display makes use of a device that decodes the available binary numbers into a low = on output, you must use the common-anode readout. If you wish to make use of a common-cathode device, IC3 must be replaced with a 7448 decoder. This newer IC will decode its binary input just like the 7447, but it will provide a high = on instead of a low = on output.

So to make life a lot easier for yourself, just stick with the schematic and parts list provided. It is not apparent at this time, but there is a reason for this madness.

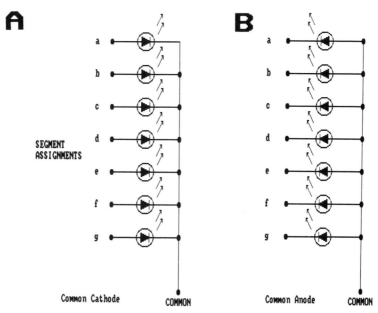

Fig. 12-3 *LED readouts come in two*
main categories: the common-cathode (A) the common-anode (B).

To end the discussion of seven-segment readouts, the LEDs also can be cascaded together to form numbers larger than one digit. Figure 12-4 shows a socket assembly available to the hobbyist at modest cost. By purchasing an assembly that will provide the correct number of spaces for your readouts, any project can truly have that professional appearance. The assembly also provides a tinted plastic window and a front-panel, mountable frame. All that remains is that each LED segment be wired to its appropriate IC output pin.

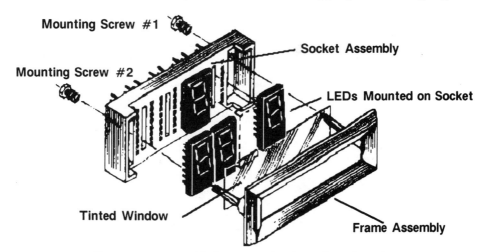

Fig. 12-4 *Preassembled mounting hardware for LED readouts*
is available from electronic mail order houses. They do cost a lot of $$$.

THE 7447 DECODER/DRIVER IC

To convert a four-bit binary number into a seven-bit output, you need some kind of decoder. You need a device that will provide a logic 0 at the needed output pins to light up the individual LED segments. It might seem like an elaborate task. With time and the understanding of logic gates, you will eventually come up with a solution. But why bother? Companies like Signetics, National Semiconductor, and others have created a device to do all these gating requirements at a blink of an eye. The IC in question is the 7447 seven-segment decoder and driver.

Figure 12-5 depicts the pinout functions of this remarkable device, and Table 12-2 shows which output pins are brought to ground while being driven by the appropriate binary code. The package accepts the standard 1-2-4-8 positive logic BCD (binary-coded decimal) when applied to pins 7, 1, 2, and 6 respectfully.

The output pins of the 7447 are capable of providing a 40 mA ground to a load (LED segment with associated dropping resistor) in the low output state and a 30 V signal in the high.

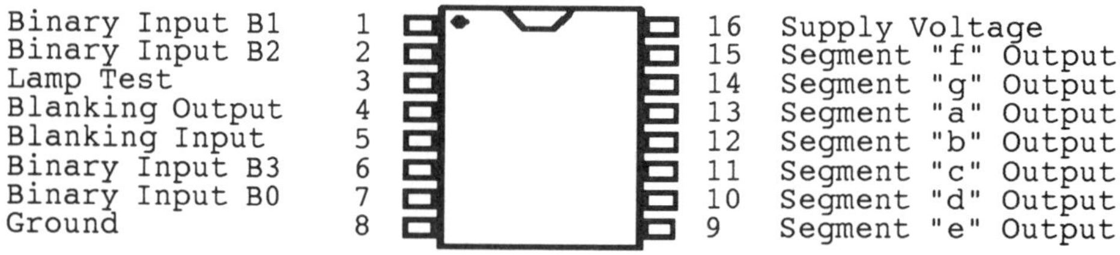

Fig. 12-5 *Pinout assignments for the 7447 decoder/driver IC.*

Table 12-2. The 7447 IC truth table.

Truth Table for the 7447 Decoder/Driver

Binary Input				Segment Output a	b	c	d	e	f	g
0	0	0	0	0	0	0	0	0	0	1
0	0	0	1	0	0	1	1	1	1	1
0	0	1	0	0	0	1	0	0	1	0
0	0	1	1	0	0	0	0	1	1	0
0	1	0	0	1	0	0	1	1	0	0
0	1	0	1	0	1	0	0	1	0	0
0	1	1	0	1	1	0	0	0	0	0
0	1	1	1	0	0	0	1	1	1	1
1	0	0	0	0	0	0	0	0	0	0
1	0	0	1	0	0	0	1	1	0	0

Note: A Logic "0" equals "ON"

Current limiting is essential for the proper operation of the device. Typically, resistors in the range of 330 Ω are wired in series with each segment output. If resistors are not used, excessive current will be drawn by the readout, thus shortening its life expectancy.

THE BINARY TO TELEPHONE CODE DILEMMA

So far, you took a look at the G8870 dual tone multifrequency receiver and the 7447 Decoder/Driver chip. If you now take these two ICs and wire the binary output of the 8870 to the binary input of the 7447, you would have a very sophisticated dialed-digit display. But there would be one major flaw. Your circuit would not be able to display the 0 number when dialed, (not the number *zero* but rather the telephone-related zero). Sound a little confusing? Focus on this problem for the next few paragraphs.

In decimal form, the number zero is meant to mean the absence of a number or the number before 1. It is represented by the binary code 0000. In telephone lingo, the number *zero* means the number 10. For instance, say you have a rotary-dial telephone and you dial for the operator. The pulsating contacts inside the dial will make and break ten times before coming to its normal resting position. But if displayed on a counting device, this device must show the number 0 (zero). (The binary code for the number 10 is 1010.)

So once again, imagine that you connect the binary output of the 8870 to the input of the 7447. If you press the tone-dial button 1, the 7447 would display the number 1. If you press the number 5 button, the 7447 would decode and display (on the seven-segment LED readout) the number 5. Now comes the flaw mentioned earlier. Say you press the operator (or 0) button of the tone dial. The G8870 IC would decode the pressing of this button into the correct binary 1010 output, but the 7447 would not understand this input. An invalid display would be produced.

To see this more clearly, please turn your attention to the conversion chart in Table 12-3. This table lists the standard binary output of a counting device if each input pulse were counted as one. As you can see, you would start off with the binary code 0000. Meaning that no pulses were received at this time. If translated back into normal, everyday English, the decimal equivalent would be zero (the number less than 1). With each successive input pulse, your imaginary counting circuit would automatically increase by one decimal count (the binary code is shown in column one of Table 12-3.)

For the sake of simplicity, say that you have inputted into the counting device, pulse number 9. The associated binary code would be 1001. So far, so good. But what if you zap your counting circuit with another pulse (pulse 10). Table 12-3 shows that the counter will automatically reset to zero (0000), where the next input pulse would be counted as pulse number 1, not pulse number 10. Binary counting such as this, where automatic reset will occur, is called binary-coded decimal or BCD for short. But for telephone circuits you need the straight binary code 1010 to represent the decimal number 10 (as you recall, this decimal 10 should be displayed on your readout as the number 0).

Table 12-3. The binary-to-decimal conversion chart.

Binary to Decimal Conversion Chart	
Binary Code	Decimal Equivalent
0 0 0 0	0
0 0 0 1	1
0 0 1 0	2
0 0 1 1	3
0 1 0 0	4
0 1 0 1	5
0 1 1 0	6
0 1 1 1	7
1 0 0 0	8
1 0 0 1	9
Reset to "0"	

It is getting complicated, but stay with it a moment longer. Take a good look at Table 12-4. This chart shows the needed binary code to display the correct decimal representation on your LED. Note that the code 0000 will now have no Display at all, and the binary 1010 will show the number 0.

You need some kind of circuit or device that will convert or change the input binary code for the number 0 (1010) into a code that the 7447 can recognize and light up the a, b, c, d, e, and f segments of the LED readout to form the representation of the number zero.

Have no fear. This reconstruction can take place in a device that can be purchased for about $1.49. It is the PROM (programmable read-only memory chip). The PROM is the subject of the next discussion.

Table 12-4. Binary-to-telephone conversion chart.

Binary to Telephone Conversion Chart	
Binary Code	Decimal Equivalent
0 0 0 0	NO Display
0 0 0 1	1
0 0 1 0	2
0 0 1 1	3
0 1 0 0	4
0 1 0 1	5
0 1 1 0	6
0 1 1 1	7
1 0 0 0	8
1 0 0 1	9
1 0 1 0	"0" or Operator

THE PROM

Advancements in electronics has made possible the ability to teach a chip binary codes. The amazing thing is that this piece of silicon will remember, for years to come, what you have taught it. You might think that memory chips are large, bulky components. But contrary to your beliefs, a 32-word (eight bits per word) electronic brain can be packed into the same amount of space that a 16-pin linear or digital integrated circuit would require. To prove this point, Fig. 12-6 illustrates how a standard PROM would look if you went out and bought one. Looks just like the G8870 or even the 7447. But its operation and wiring requirements are quite different.

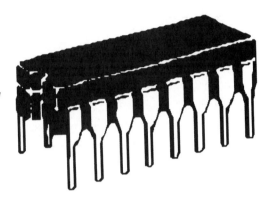

Fig. 12-6 *An electronic memory can come in a standard 16-pin DIP (dual inline package).*

First of all, you need some kind of device that can teach your PROM the binary codes that are needed. This device is called a programmer, and it just so happens that in appendix B, there are three programmers that you can choose from. Whichever programmer you decide to build will be needed for this project and a few more advanced circuits presented later.

See the importance of this chapter? You need the programmer to teach the electronic memory (74188 or 8223) the codes that your future talking telephone and the talking telephone ringer will need for operation.

So for now, flip ahead a few chapters and look over the programmers. Build one. Learn how the memory chip works and how to program it. Once you have mastered the basics of programming and you are confident that your programmer is operating correctly, come back and continue with the discussion on the dialed-digit display.

IC PROGRAMMING

Now that you have an operational PROM programmer you need some sort of binary table to help us convert the standard output code of the G8870 into a nonstandard binary output.

To determine the correct binary input for the 7447, put in the first column of a piece of paper, all the possible binary outputs from the G8870. Of course, the digit display does not make use of the # or the ∗ button, so they are eliminated from this list. This leaves the usual 0 (zero) to 0 (operator) buttons. In the second column, write down the code you wish at the output of the PROM chip. In the third column, indicate the decimal value of each code. To save you time, Table 12-5 is a list as described above.

Table 12-5. Binary-to-telephone conversion chart using the 74188 memory chip.

Binary to Telephone Conversion Using 74188 Memory Chip		
Binary Code	Memory Chip Conversion	Decimal Equivalent
0 0 0 0	0 0 0 0	0
0 0 0 1	0 0 0 1	1
0 0 1 0	0 0 1 0	2
0 0 1 1	0 0 1 1	3
0 1 0 0	0 1 0 0	4
0 1 0 1	0 1 0 1	5
0 1 1 0	0 1 1 0	6
0 1 1 1	0 1 1 1	7
1 0 0 0	1 0 0 0	8
1 0 0 1	1 0 0 1	9
1 0 1 0	0 0 0 0	"0" or Operator

Cannot convert any standard Binary Code to deliver a "No Display" level so use the "STROBE" pin of the G8870 IC. This low will be delivered to pin 4 of the 7447 IC. The low will "BLANK" out all characters until a valid tone signal is received, then display the dialed number.

The first column indicates the binary code for the numbers zero to ten. Column two shows the codes you wish to be present at the output of the 74188 PROM chip. The third is the decimal-equivalent column.

From binary 0000 to 1001, the output of the 74188 is the same as the coded output of the 8870. So coming up with the proper truth table for these ten numbers is not a ticklish project. But wait—programming will, without a doubt, become extremely difficult in later chapters.

For now, consider what happens when the input binary code equals 1010. With this input, the decoder/driver (7447) must display a 0 on the LED readout. The only way this can be done is to apply four zeros at the chips input (0000). The required output will be the same for an input of 0000 and 1010, but you do have an ace in the hole. The ace is the strobe line (pin 15) of the 8870. If you take this strobe output and connect it to pin 5 of the 7447 (blanking input), you can create a situation that will prevent a No-Digit display only when no tone-dial buttons have been pressed. In reality, the binary 0000 is present at the data input pins of the 7447 but you inhibited or prevented the normal operation of the chip (7447) by placing a logic low on the blanking-input pin.

When the pressing of the operator key is detected, the PROM will generate the needed 0000 binary output, just like the decimal 0 input did, but with one big difference. The strobe line of the 8870 is now high, indicating the input of a valid DTMF tone. At this time, the blanking input pin is high. This high will activate or turn on the 7447. The high will allow the chip to display the decoded decimal number. In the case of the 1010 input, the readout will display the required 0 number.

With this long and drawn-out explanation behind you, turn your attention to the programming of an actual truth table. Table 12-6 illustrates a listing with three columns. The first shows the decimal display output if you have assembled the programmer making use of the digital thumbwheel switches. The second column indicates the position of switches SW3 to SW5 if you have assembled the programmer making use of the user-defined address. And finally, the third column indicates the eight SW8 switch settings used on both programmers.

Just remember that if you make use of the SW3 to SW7 switches, make sure that you double check all switch positions before you press the program button (SW1). If you make a mistake at this point, there is no turning back. You cannot reprogram an error. So double check all switch settings beforehand.

Table 12-6. To teach the PROM (74188 chip) how to produce a standard LED display, you must program this information into the chip. Here is the truth table that shows you how.

Truth Table for Programming the 74188

Thumbwheel Display	Binary Code Input					Memory Chip Programmer SW8 Setting							
	SW7	SW6	SW5	SW4	SW3	B7	B6	B5	B4	B3	B2	B1	B0
00	0	0	0	0	0	0	0	0	0	0	0	0	0
01	0	0	0	0	1	0	0	0	0	0	0	0	1
02	0	0	0	1	0	0	0	0	0	0	0	1	0
03	0	0	0	1	1	0	0	0	0	0	0	1	1
04	0	0	1	0	0	0	0	0	0	0	1	0	0
05	0	0	1	0	1	0	0	0	0	0	1	0	1
06	0	0	1	1	0	0	0	0	0	0	1	1	0
07	0	0	1	1	1	0	0	0	0	0	1	1	1
08	0	1	0	0	0	0	0	0	0	1	0	0	0
09	0	1	0	0	1	0	0	0	0	1	0	0	1
10	0	1	0	1	0	0	0	0	0	0	0	0	0

To program a 74188 Memory Chip a "PROGRAMMER" must be constructed. See Appendix B for three devices used to "teach" your IC the above table.

LOGIC "1"

Switch Settings  SW3 to SW7

LOGIC "0"

THE PROGRAMMING OF THE 74188 PROM

Do you think that this book would just give you an explanation on programming techniques and leave you there to fend for yourself? No way. Below is a step-by-step procedure for programming the 74188 (or 8223) per the truth table presented in Table 12-6. Just follow each line very carefully. Recheck all switch settings, and take your time.

By following these precautions now and in the future, you will prevent the unnecessary mistakes that will force you to discard a mis-programmed $1.49 chip. Mistakes of this type can add up.

Programming Procedure

1. Seeing that a thumbwheel display of 00 and a decimal output of 0 require a logic 0 from B0 to B7, you can skip this program step.
2. Flip the thumbwheel to display a 01 or flip SW3 to its UP position. Make sure SW2 is in the program position. Then adjust SW8 to its B0 position. When all is OK, press SW1 (the program switch). The pressing of this switch *blows* out the internal fuse of the 74188 IC at address 01 position B0. To verify that this fuse is blown, just flip SW2 to its verify position. The programmer LED will light indicating a logic 1 at this address.
3. Flip the thumbwheel to display a 02 or flip SW4 to its UP position (make sure that SW3 from the previous step is returned to the logic 0 position and SW2 is in program). Adjust SW8 to its B1 position. Then press SW1 to program. Flip SW2 to the verify position to check for a logic 1 at this address. Flip all switches to their Logic "0" position and SW2 to PROGRAM.

To avoid being redundant, the following is omitted from the remaining programming procedure:

- Verify all programming addresses by flipping SW2 to its verify position; then return it to program.
- When finished programming an address, remember to flip SW3, SW4, SW5, SW6, and SW7 to the logic 0 position.

4. Flip the thumbwheel to display a 03 or flip SW3 and SW4 to the UP position. Set SW8 to the B0 position, then press SW1 to program. With SW3 and SW4 in the same positions, readjust SW8 to its B1 position. When positioned, press SW1 to program.
5. Flip the thumbwheel to the 04 position or flip SW5 to its up position. SW8 to B2. Press SW1 to program.
6. Flip the thumbwheel to the 05 position or flip SW3 and SW5 to its up position. Adjust SW8 to B0. Press SW1 to program.
7. Still at binary 5 address, readjust SW8 to its B2 position. Press SW1 to program.
8. Flip the thumbwheel to the 06 position or flip SW4 and SW5 to the up position. Adjust SW8 to B1. Press SW1 to program.
9. Still at address 6, readjust SW8 to its B2 position. Press SW1 to program.
10. Flip the thumbwheel to the 07 position or flip SW3, SW4, and SW5 to the up position. Adjust SW8 to B0. Press SW1 to program.

11. Still at address 7, readjust SW8 to the B1 position. Press SW1 to program.
12. Still at address 7, readjust SW8 to its B2 position. Press SW1 to program.
13. Flip the thumbwheel to the 08 position or flip SW6 to its up position. Adjust SW8 to its B3 position. Press SW1 to program.
14. Flip the thumbwheel to the 09 position or flip SW3 and SW6 to the up position. Adjust SW8 to B0. Press SW1 to program.
15. Still at address 9, readjust SW8 to its B3 position. Press SW1 to program.
16. With the thumbwheel displaying 10 or SW4 and SW6 in the up position, the truth table indicates that there are no SW8 settings.

So this ends our first programming session.

THE CONSTRUCTION OF THE DIALED-DIGIT DISPLAY

Did you think you would never get to the construction of the digit display because the PROM explanation was so long? It did serve its purpose and you will need the programmer and its theory of operation in the following chapters of this book. So for now, turn your attention to the building of the digit display.

By referring to the schematic of the circuit in Fig. 12-1, you can see that the construction of the display is not at all complicated. Seeing that you now have a programmed memory chip, complexity of the design has been cut dramatically. The circuit itself can be constructed on a perf board or if you wish, you might want to design your own printed circuit board.

As usual, whether you use a perf or PC board, some basic building rules must be observed if you want the circuit to be both operational and sturdily assembled. If the digit display is one of your first tries in building an electronic circuit, IC sockets might be a good investment. By investing about a dollar or two for IC sockets, the lives of your expensive integrated circuits might be spared the destructive heat of soldering. Hobbyists have the idea stuck in their head that a lot of heat is required in order to make a properly soldered joint. This idea is by no means true. Fast, accurate soldering is the key to successful kit building. A one-second soldering job can create a joint just as clean and just as perfect as if the iron were heating the connection for fifteen seconds.

So before starting any assembly project, practice soldering on wire scraps. Then when you are confident in your ability, and only then, take the plunge by trying your skills on a project. Even then, consider IC and transistor sockets instead of soldering.

Except for the three ICs and the seven-segment LED, there are no components with polarities, which makes the digit display an ideal beginner's project. But take care when choosing the readout. Remember from previous discussions, the seven-segment LED comes in two configurations: the common anode and the common cathode. The circuit, as it stands, requires a common anode readout. This allows you to light individual segments by applying a low or ground to any of the a-to-g pins. The common connection pin of the readout must be connected to a +5Vdc. If you purchase a common cathode, do not expect the circuit to operate. It will just sit there very quietly and not do a thing. It just will not work.

The dialed-digit display should not generate any difficulties except for the purchasing of the G8870. This IC cannot be found at your local Radio Shack store. This is a special integrated circuit. This is the reason, Del-Phone Industries of Spring Hill, Florida, has agreed to sell the G8870 in quantities of one or more. So drop them a line and request the current pricing and availability of this Teltone chip.

POWER SUPPLY

For the proper operation of the circuit, a power supply delivering 5 Vdc is needed. This supply can be designed with a 6 to 9 Vac wall transformer. But make sure that the ac power is properly rectified and filtered. If this scheme is a little over your head, you might want to consider using a dc Wall (5 to 6 Vdc) transformer with a current rating of about 250 mA. This inexpensive device can provide the proper voltage and current without brain strain on your part.

INSTALLATION

Just as the previous digit display was connected in parallel with the telephone, so it is with this circuit. You might want to connect the circuit input (C1 and C2) to the screw terminals of a telephone wall jack. But to make a quick and accurate installation, invest a few dollars and purchase a TEE adapter. This adapter will allow you to easily connect two telephone related devices to one modular wall jack.

When your display is properly hooked up in parallel with the telephone line, turn on the juice. Lift a telephone receiver that uses a tone dial. Press one button at a time, and notice the display on the readout. If it is operating properly, the LED should display the decimal number of the button you have pressed. Now, conduct the same test as described above for all remaining keypad digits. While testing, take special notice when the operator button is pushed. The LED should display a 0. If it does, you are in business. If not, disconnect power and retest the programmed binary code in the 74188 IC. There might be a mistaken address or output bit.

You should also notice the no-digit display when no tone dial buttons have been pressed. The readout should be blank. If not, check that the G8870 Strobe is connected properly to the 7447 IC.

With all electrical testing complete, it is time to install your project in an attractive housing. Whether it's a metal or plastic box, a rectangular cutout must be made in the front panel. This cutout will allow you to see the readout. If you wish, you can also install a piece of *smoked* (tinted) plastic behind the cutout. This plastic was specially molded with a tinted or smoked look to it. The tint prevents you from seeing the physical outline of the readout. But when lit by the circuit, the segment light will penetrate the plastic and provide a very striking display. Tinted plastic panels can be purchased from Radio Shack and other fine electronic stores. An example of this can be seen in Fig. 12-4. This figure makes use of a four socket LED assembly, but you can modify this arrangement to accommodate a single seven-Segment Readout that is being used on your new dialed-digit display.

PARTS LIST FOR THE DIALED-DIGIT DISPLAY (SEVEN-SEGMENT LED OUTPUT)

R1 R2 R4	100KΩ resistors
R3	47KΩ resistor
R5	50KΩ 5 percent resistor
R6	220KΩ resistor
R7 R8 R9 R10	1KΩ resistor
R11 to R17	330 Ω resistor
C1 C2	.01 μF disk capacitor
C3	.1 μF disk capacitor
IC1	G8870 DTMF receiver IC*
IC2	74188 programmed PROM IC
IC3	7447 decoder/driver IC
XTAL1	3.5 MHz crystal
LED1	Common-anode seven-segment readout
1	Printed circuit or perf board
1	Telephone line cord
1	5 Vdc power supply
1	Housing

*Available from Del-Phone Industries, P.O. Box 5835, Spring Hill, Florida 34606.

☎13
Dialed-Digit Display with PROM (Programmable Read-Only Memory) Character Generator

CHAPTERS 11 AND 12 INCLUDE THE CONSTRUCTION OF DIGIT DISPLAYS USING READILY available components like the 7447 and the seven-segment LED readout. Chapter 13 takes the concept just a bit further by eliminating the use of the 7447 IC altogether and substituting a PROM memory chip. You did something like that in chapter 12, right? But this time you will simplify the final circuit by taking a 74188 or 8223 memory chip and programming it to act as if it were a *character generator*.

Character generator is a new buzz word to learn? Yes, but it is an important concept to understand. Read on to see how you can put this technology to work.

A character generator is a device that will accept a standard binary code as its input and, as programmed by you, illuminate selected segments of an LED readout. The lighting of these individual sections will then represent the decimal equivalent of the input. In other words, if you place the binary number 0100 (decimal four), at the input, the PROM will sense this code and produce at its output pins (B0 to B7), a unique code. When the output is connected to an LED readout, it will display the number four. This is done by lighting the correct four segment bars that make up the number.

To see which of the seven bars are illuminated to produce the numbers between 0 and 9, turn back to Table 12-1. This table will not only show you the segments needed to produce the numbers but it also indicates the segment lettering (a, b, c, d, e, f, and g) that must be taken into consideration if you wish to create proper numbering.

Just how to illuminate an LED segment using a character generator is the

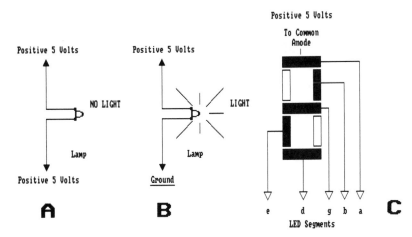

Fig. 13-1   *It is easy to make an LED readout work.*

next problem. For now, turn your attention, for just a moment, to Fig. 13-1. Represented here are three illustrations. Figure 13-1A shows a bulb with an applied voltage of 5 V to the top half. The bottom half of the same bulb also has a voltage of 5 V. The drawing shows that the lamp will not light because there is no difference of *potential* (difference in voltage) between these two points. Figure 13-1B shows what would happen if you take the bottom section of the bulb and connect it to the common ground of the same 5 V power supply. Now see that there is a 5 V voltage on one side of the component and a ground at the other end. There is now a path for electrons to follow or there is now a difference of potential between these two points of about 5 V. It is this difference of potential that transforms electrical energy into visible light in the bulb.

This principle can be taken one step further, as displayed in Fig. 13-1C. By applying a positive 5 V to the common connection point of a common-anode LED readout, you can light any one of the seven available segments by allowing one continuous path for the electrical energy to follow. And this can be done by applying a ground to the desired segment.

Figure 13-1C illustrates this quite well. As you can see, if you apply a ground at segments a, b, d, e, and g, you can create the number 2. If you apply grounds at other bar segments, all of the numbers from 0 (zero) to 9 can be produced. You can even create a number of alphabet characters using this scheme. But more on that a bit later.

How does this apply to your character generator? More than you realize at this point. So please continue.

If you take a brand-new 74188 PROM chip and put it on one of the programmers illustrated in appendix B, you will quickly see that all addresses produce a logic 0 output at pins B0 to B7. Take an LED readout and connect segment letter a to the B0 output of the 74188, segment b to the B1 output, segment c to B2 output, and so on. You have the makings of an inexpensive character generator. But you will notice that when the wiring is complete, the 74188 provides eight separate output points and our readout uses only seven.

That is no problem. Just place the B7 output to ground and forget about it.

If you apply the needed 5 V to your experimental circuit, you will see that all of the seven segments of the readout will light. Why? Because there is a difference of potential between the common-anode connection point and the segment bars. And a difference of potential will allow the flow of electrons in the wire.

So what do you do if you do not want segments to light? Using a PROM Programmer (see appendix B), you can blow out one of the 74188 internal fuses and create a logic 1 at that address location. If a logic 1 (or 5 V level) is applied to one of the segments, the difference of potential is eliminated, thus preventing the illumination of that particular segment.

So if you program selected logic 1s into your memory at selected binary input codes, you can create a memory that will display whatever number, letter, or character that you desire.

PROGRAMMING THE CHARACTER GENERATOR

Table 13-1 shows the truth table that you will use to program the character generator. For an example on how the truth table was designed, take the binary code 0000 and use it as an input to the PROM chip. If you recall from chapter 12, the number zero represents the absence of something or the number less than one. The text in chapter 12 called this a no-display output. Now with a programmed character generator, you can easily prevent any display from appearing just by programming the memory with a series of seven logic 1s at the output locations B0, B1, B2, B3, B4, B5, and B6. Logic 1s at these locations

**Table 13-1. You need to know what to teach
your PROM chip. This is the truth table for the program.**

Truth Table for the PROM Character Generator (Common Anode)													
Thumbwheel Display	Binary Code Input SW7 SW6 SW5 SW4 SW3					Memory Chip Programmer SW8 Setting B7 B6 B5 B4 B3 B2 B1 B0							
	SW7	SW6	SW5	SW4	SW3	B7 N/C	B6 g	B5 f	B4 e	B3 d	B2 c	B1 b	B0 a
00	0	0	0	0	0	0	1	1	1	1	1	1	1
01	0	0	0	0	1	0	1	1	1	1	0	0	1
02	0	0	0	1	0	0	0	1	1	0	0	1	0
03	0	0	0	1	1	0	0	1	1	0	0	0	0
04	0	0	1	0	0	0	0	0	1	1	0	0	1
05	0	0	1	0	1	0	0	0	1	0	0	1	0
06	0	0	1	1	0	0	0	0	0	0	0	1	0
07	0	0	1	1	1	0	1	1	1	1	0	0	0
08	0	1	0	0	0	0	0	0	0	0	0	0	0
09	0	1	0	0	1	0	0	0	1	0	0	0	0
10	0	1	0	1	0	0	1	0	0	0	0	0	0

To program a 74188 Momory Chip a "PROGRAMMER" must be constructed.
See Appendix B for three devices used to "teach" your IC the above table.
NOTE: PROM Output (B7) is not used with this application.

```
                              LOGIC "1"
                              ▓▓▓▓▓
Switch Settings               ▓▓▓▓▓        SW3 to SW7
                              □□□□□
                              LOGIC "0"
```

will prevent any of the seven segments from lighting up, thus creating a no-display output.

Let's take one more example. Remember in chapter 12 that if you press the operator button of a tone-dial telephone, the G8870 will output the binary code 1010. This will represent the number 0. So to display a 0 on the readout, just change the binary code address (SW3 to SW7 or set the programmer thumbwheel to display a 10) to 1010 and blow the PROMs fuse at location B6.

When properly connected to the readout of the finished project, segments a, b, c, d, e, and f will have a ground placed on them, and segment g will have a logic 1. A ground (or logic 0) will light the segments, and a logic 1 will keep that segment off. The resulting display from this example will be the number 0.

The programming method for the remaining digits will be basically the same except that the segments being illuminated will be different.

Programming seems to be a complicated concept to grasp at first. This is only your second try at it, so let's do it together. Get your programmer warmed up, plug in a new 74188 PROM. Carefully read and then reread the following step-by-step procedure for programming your first character generator.

1. Rotate the thumbwheel to display the number 00 or flip SW3, SW4, SW5, SW6, and SW7 to logic 0 positions. Make sure SW2 is in program. Then rotate SW8 to its B0 position. When all is set, press SW1 (the program switch). By pressing this switch, the internal fuse located at position B0 at address 00000 will be blown. To verify that this fuse has been eliminated, just flip SW2 to its verify position. The programmer LED will light, indicating a logic 1 level at this address.

   From this point on, the following is omitted from the programming procedure, but it should be done anyway.

   1A. Verify all programming addresses by flipping SW2 to its verify position; then return it to program.

   1B. When programming is complete for each address, remember to flip SW3, SW4, SW5, SW6, and SW7 to their logic 0 positions, otherwise incorrect data will be programmed.

2. With the thumbwheel still in the 00 position or SW3, SW4, SW5, SW6, and SW7 in logic 0 positions, do the following:

   2A. Rotate SW8 to its B1 position. Then press SW1 to program. Test for a logic 1 by flipping SW2 to verify.

   2B. Rotate SW8 to its B2 position. Press SW1. Then verify for a logic 1 level by flipping SW2 and checking LED.

   2C. Rotate SW8 to its B3 position. Press SW1. Then verify for a logic 1 level by flipping SW2 and checking LED.

   2D. Rotate SW8 to its B4 position. Press SW1. Then verify for a logic 1 level by flipping SW2 and checking LED.

   2E. Rotate SW8 to its B5 position. Press SW1. Then verify for a logic 1 level by flipping SW2 and checking LED.

   2F. Rotate SW8 to its B6 position. Press SW1. Then verify for a logic 1 level by flipping SW2 and checking LED.

3. Rotate the thumbwheel to display a 01 or flip SW3 to its up position.

Then do the following:

3A. Rotate SW8 to its B0 position. Then press SW1 to program.
Test for a logic 1 by flipping SW2 to verify.

3B. Rotate SW8 to its B3 position. Then press SW1 to program.
Test for a logic 1 by flipping SW2 to verify.

3C. Rotate SW8 to its B4 position. Then press SW1 to program.
Test for a logic 1 by flipping SW2 to verify.

3D. Rotate SW8 to its B5 position. Then press SW1 to program.
Test for a logic 1 by flipping SW2 to verify.

3E. Rotate SW8 to its B6 position. Then press SW1 to program.
Test for a logic 1 by flipping SW2 to verify.

4. Rotate the thumbwheel to display a 02 or flip SW4 to its UP position.
Then do the following:

4A. Rotate SW8 to its B2 position. Then press SW1 to program.
Test for a logic 1 by flipping SW2 to verify.

4B. Rotate SW5 to its B5 position. Then press SW1 to program.
Test for a logic 1 by flipping SW2 to verify.

5. Rotate the thumbwheel to display a 03 or flip SW3 and SW4 to their
up position. Then do the following:

5A. Rotate SW8 to its B4 position. Then press SW1 to program.
Test for a logic 1 by flipping SW2 to verify.

5B. Rotate SW8 to its B5 position. Then press SW1 to program.
Test for a logic 1 by flipping SW2 to verify.

6. Rotate the thumbwheel to display a 04 or flip SW5 to its UP position.
Then do the following:

6A. Rotate SW8 to its B0 position. Then press SW1 to program.
Test for a logic 1 by flipping SW2 to verify.

6B. Rotate SW8 to its B3 position. Then press SW1 to program.
Test for a logic 1 by flipping SW2 to verify.

6C. Rotate SW8 to its B4 position. Then press SW1 to program.
Test for a logic 1 by flipping SW2 to verify.

7. Rotate the thumbwheel to display a 05 or flip SW3 and SW5 to its up
position. Then do the following:

7A. Rotate SW8 to its B1 position. Then press SW1 to program.
Test for a logic 1 by flipping SW2 to verify.

7B. Rotate SW8 to its B4 position. Then press SW1 to program.
Test for a logic 1 by flipping SW2 to verify.

8. Rotate the thumbwheel to display a 06 or flip SW4 and SW4 to its up
position. Then rotate SW8 to its B1 position. Press SW1 to program.
Test for a logic 1 by flipping SW2 to verify.

9. Rotate the thumbwheel to display a 07 or flip SW3, SW4 and SW5 to
their up positions. Then do the following:

9A. Rotate SW8 to its B3 position. Then press SW1 to program.
Test for a logic 1 by flipping SW2 to verify.

9B. Rotate SW8 to its B4 position. Then press SW1 to program.
Test for a logic 1 by flipping SW2 to verify.

9C. Rotate SW8 to its B5 position. Then press SW1 to program. Test for a logic 1 by flipping SW2 to verify.

9D. Rotate SW8 to its B6 position. Then press SW1 to program. Test for a logic 1 by flipping SW2 to verify.

10. Number 8 programming—seeing that you want all seven-segments of the LED to light at this address, do not program any B outputs at this time. Go to step 11.

11. Rotate the thumbwheel to display a 09 or flip SW3 and SW6 to their up positions. Then rotate SW8 to its B4 position. Press SW1 to program. Test for a logic 1 by flipping SW2 to verify.

12. Rotate the thumbwheel to display a 10 or flip SW4 and SW6 to their up position. Then rotate SW8 to its B6 position. Press SW1 to program. Test for a logic 1 by flipping SW2 to verify.

Once you have completed the programming of the memory, retest all addresses and outputs for the proper logic levels. If you have inadvertently programmed a logic 1 where a logic 0 must reside or vice versa, you will immediately see the problem on the readout. It will display an erroneous character or shape. If the difficulty lies with a logic 0 output where a logic 1 should have resided, there is no problem to correct the error; just blow the correct fuse. But if you programmed a logic 1 where a logic 0 should have been, it cannot be repaired. The IC must be thrown away, and you must start the programming procedure all over again. That is the reason you should recheck all switch settings before hitting that SW1 button; there is no turning back.

THE CIRCUIT

Now that you have a truth table burned into chip memory, you have to do something with it. And so you shall. By turning your attention to Fig. 13-2, you will find the complete schematic for the dialed-digit display. By comparing Fig. 13-2 with the schematics of chapters 11 and 12, you can see that by using a memory chip to remember a table of 0s and 1s, you can greatly reduce the complexities of the digit display. This also holds true for any circuit that requires memory of some sort. So if you come across a circuit or if you have an idea for a fun-time project, keep in mind that 74188s cost only $1.49 each, and that they can be programmed without any hassle whatsoever.

Through your inspection of Fig. 13-2, you can see that the G8870 DTMF IC is used in this project. So an in-depth discussion explaining the operation of the chip is not really necessary here. Just keep in mind that the G8870 converts the tones generated by a telephone TTP (touch-tone pad) into its corresponding binary equivalents. The output can be found at pins 11, 12, 13, and 14 (D0, D1, D2, and D4 respectively).

When any TTP button is pressed, a binary code is made available at the output of the G8870. In order to have a 0000 binary when no buttons are pressed, you must do some fancy footwork. This is how you can accomplish this.

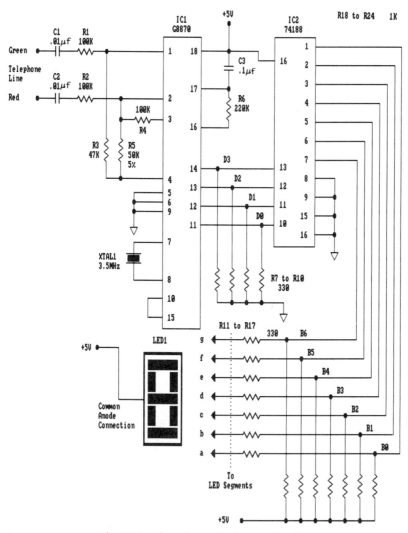

Fig. 13-2 *The schematic diagram for the dialed-digit display with a programmed character generator.*

Pin 15 of the DTMF IC is the strobe. It will give a logic 1 only when a TTP button is pressed. Pin 10 is a pin that enables the output data. If this pin were made high, the binary outputs (D0 to D3) will be made present. If the enable pin were made low, no output would be present. If you combine the strobe output and the enable input, you can easily control the presence and absence of data. By pressing a button, the strobe goes to logic 1. The enable needs a logic 1 to provide data to the output of the chip. So in reverse, if no button were pressed, the strobe pin will be low (or logic 0). A low at this pin will disable any data output of the G8870.

There still remains one difficulty to overcome—the output of the DTMF IC is made to provide what is called a *tristate output*. A device making use of this technology can provide three different output stages for the IC. Stage one is a logic 1, stage two is a logic 0, and stage three is a high-impedance (or open-circuit) output. This design can complicate matters a bit, because if there is a low at the enable pin, the output data will be at stage three or at its high-impedance state. But the programmed 74188 needs four logic 0s inputs to indicate a no-display readout, not an open circuit.

Here is a little secret for you. To overcome the tristate output of this or any other IC, just take four resistors (R7 to R10) and place them from all data input lines (G8870 pins 11, 12, 13, and 14) to ground. Then when the device is at a stage three state, the PROM memory will see four grounds (or 0000 at its binary input). But if a logic 1 is present at the output of the DTMF IC, the resistors will have no effect on the circuit because the value of resistors R7 to R10 were chosen to provide either a 330 Ω to ground path, or they will allow a +5 V logic level to pass without any appreciable losses in signal strength.

So with the circuit wired as described as above, there will be no display on the readout when no TTP buttons are pressed. The number corresponding to the dialed digit will be displayed only when the button is pressed.

Resistors R18 to R24 are standard pull-up resistors, and they provide the needed voltage for the PROM chip. In contrast, resistors R11 to R17 are used for current limiting. If you recall, LEDs will gobble up as much current as you are willing to feed them. So to prevent the destruction of the readout, resistors are placed in series with all segment lines. Do not try to place one resistor in the common-anode connection. It is just not an acceptable practice, and it should be avoided.

CONSTRUCTION

Just as with the other projects in this book, a PC and perf board are the best construction media available. Although making a PC board is expensive and time consuming, it is by far your best choice.

Perf boards do provide a flexibility that PCs do not. Concentrate on the assembly because mis-wires and solder bridges can spell havoc for expensive and delicate integrated circuits.

Whichever medium you choose, just keep in mind that ICs (and LED readouts) require proper orientation in the circuit. Otherwise the operation of a circuit, when complete, is just a pipe dream. As for the construction of the dialed-digit display, the only devices you need be concerned with are IC1, IC2 (it is a good idea to make use of a socket for IC2, just in case you find a programming error on the chip) and the LED readout. If you take care in the wiring of these components, there is no way you will encounter any difficulty with the finished project.

TESTING AND INSTALLATION

Before applying the needed 5 V power, plug in your programmed memory chip. Pay close attention to its proper orientation in the socket.

The next requirement is to connect the device to the telephone line. This is accomplished by using a standard modular telephone line cord, by connecting the red and green (spade-tipped ends) to the digit display where indicated. Then snap the connector into a wall jack, and installation of your circuit is easily and painlessly made.

When complete, apply the proper voltage to the board. Lifting the telephone headset, press three or four buttons (one at a time) and take notice to the LED readout. The reconstructed numeral display should correspond to the button being pressed at that time. This test should be conducted for the 10 buttons of the tone dial (The # and the * will display the number 8 on the readout unless you program the circuit to produce some other character. Have you an idea on how to expand the circuit a little further?)

FINAL TOUCHES

Once you have tested the dialed-digit display and are satisfied with its operation, the next thing to consider is the purchase of a housing. Consider plastic or metal housings when applying that final touch to any project. To help beautify your housing or cabinet of your choice, Fig. 13-3 has a template of a suggested front-panel layout.

This template can be copied on a copy machine from the book, cut down to size, and glued in place. Front-panel templates can give that special project a little extra TLC (No, this is not a new buzz word). And this extra TLC will be shown to all your friends and neighbors to whom you show off your new dialed digit display.

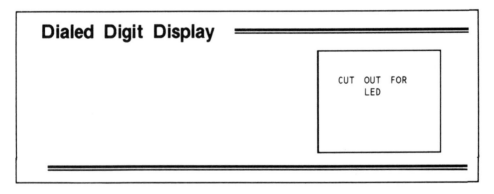

Dialed Digit Display

CUT OUT FOR LED

Fig. 13-3 *The template for the complete digit display.*

A FINAL NOTE

In the beginning of this chapter, you read that an LED readout can display a number of alphabet characters. Table 13-2 illustrates just that. By illuminating certain segments, an LED readout can be made to display the letter *A*, *F*, and so on. But there is one difficulty with this scheme. Decoder/driver ICs like

Table 13-2. By programming your PROM, you can display
characters other than the normal 0 to 9 numbers. Alphabet letters are also possible.

Additional LED Characters		
Character	**Segment Assignment**	**Display**
#'s 0 to 9	See Table 12-1	
A	a,b,c,e,f,g	R
C	a,d,e,f	C
c	d,e,g	c
E	a,d,e,f,g	E
F	a,e,f,g	F
H	a,b,e,f,g	H
I	a,b	I
J	b,c,d,e	J
L	d,e,f	L
O	a,b,c,d,e,f	O
P	a,b,e,f,g	P
S	a,c,d,f	S
U	b,c,d,e,f	U

the 7447 or the 7448 just decode the numbers 0 to 9. They do not even consider
any alphabet characters. But you can. How? By programming the desired
alphabet characters onto your PROM. It is that easy.

As the circuit stands now, if you press the # or * button of a telephone
TTP, an 8 will be displayed. That is because the binary code associated with
the * (1011) and the # (1100) have not been given any logic 1s. All are 0, and a 0
will provide that (buzz word again) difference of potential. If you combine the
dialed-digit display with the hi-tech MOH (chapter 9) you can have the LED
display an H (hold) when the * is pressed. And have the readout show the letter
C (connect) when the # is pushed.

Now you have the building blocks to expand your curiosity into the realm
of memory chips and digital electronics. Take this thirst for knowledge and
quench it. Quench it by designing, constructing, and creating electronic
devices that would stun the ingenuity and imagination of engineers only five
years ago.

PARTS LIST FOR THE DIALED-DIGIT DISPLAY (CHARACTER GENERATOR)

R1 R2 R4	100kΩ resistors
R3	47kΩ resistor
R5	50kΩ 5 percent resistor
R6	220kΩ resistor

R7 to R10	330 Ω resistors
R11 to R17	330 Ω resistors
R18 to R24	1 KΩ resistor
C1 C2	0.01 μF disk capacitor
C3	0.1 μF disk capacitor
IC1	G8870 DTMF receiver
IC2	74188 PROM (programmed—see text)
XTAL1	3.5 MHz crystal
LED1	Standard seven-segment LED
1	Telephone line cord
1	Housing
1	PC or perf board

☎14

Digital Telephone Lock*

UNAUTHORIZED TELEPHONE CALLS CAN ROB YOU BLIND. WITH YOUR HOME telephone you might think that you have control over the amount of the monthly bill, but in reality, you do not. If you have a business telephone system, it is very difficult to police the incoming and outgoing calls of employees.

Unauthorized telephone costs are usually discovered only if or when the phone company sends you an itemized bill that shows the number of local and long-distance calls that were made. Although itemization gives you proof that calls were made, it does not tell you who made them.

The only way to prevent the unwarranted use of a home or business telephone is to eliminate the ability to make the calls by removing all instruments from the premises or by adding a security system in the form of an invisible electronic lock—one that can only be opened by a special user-programmed, four-digit code. The code can be entered through a tone-dial telephone keypad.

THE TELE-GUARD II

Just such a telephone security device is the electronic combination lock, which for simplicity throughout this chapter will be called Tele-Guard II (see Fig. 14-1). The easy-to-build Tele-Guard, which can be installed into practically any single-line tone-dial telephone system, protects your telephone line from unauthorized use by actually disconnecting the device from the line until a four-digit code is entered. The code causes a small relay inside the Tele-Guard to close and connect the telephone to the outside line.

You might ask if the unlocking code must be entered if a call is being received. The answer is no! Tele-Guard II automatically bypasses its electronic lock so that the phone can both ring and be answered without entering any code at all, as if the Tele-Guard did not exist. When the conversation has been terminated, Tele-Guard will automatically rearm the locking device.

---

*Reprinted with permission from October/November 1988 Radio/Electronics ©Copyright 1988 Gernsback Publications

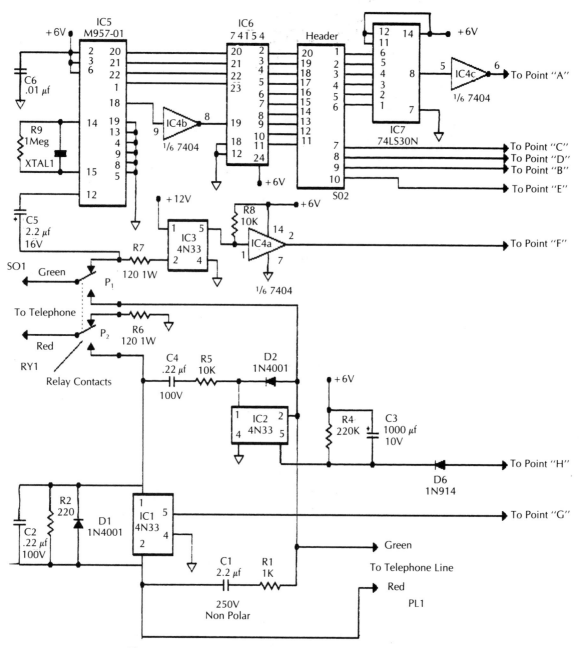

Fig. 14-1A *Part 1 of the schematic for the digital telephone lock.*
(Reprinted with permission from Radio Electronics Magazine *October 1988 & November 1988 issues.*
© *copyright Gernsback Publications Inc. 1988.)*

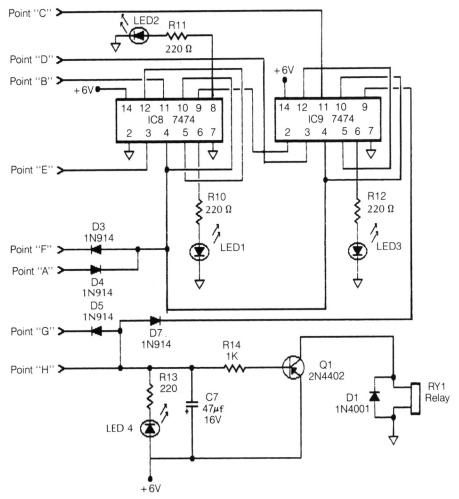

Fig. 14-1B *Part 2 of the schematic for the digital telephone lock.
(Reprinted with permission from* Radio-Electronics Magazine *October 1988 &
November 1988 issues.
© copyright Gernsback Publications Inc. 1988.)*

DIGITAL PROTECTION

If you know how many digits are required to unlock a device, most keypad operated security can eventually be broken by entering every possible combination from the dial pad. Devices such as this do exist. You might have seen one on the television show *Mission Impossible*, where a small hand-held computer generates numbers until the correct code is eventually entered, thus opening up the safe containing a kings ransom in diamonds or gold.

But with the digital telephone lock, you are not talking about protecting gold or diamonds from theft, but Tele-Guard II does have an automatic reset feature built in. The locking circuitry inside Tele-Guard will reset every time an incorrect digit is entered into the system. That means that even if the first two digits of the code are guessed correctly by luck, if the third number is incorrect, Tele-Guard senses the entry error and resets the complete circuit. It wipes out all memory of the first two correctly entered digits, thereby requiring that the first two numbers be entered correctly again, followed by the correct third and eventually by the fourth correct digit. This security design provides the highest possible protection for any type of tone dial telephone.

DTMF IS THE KEY

As mentioned in earlier chapters, tone-dial telephones use a special kind of signalling called DTMF. In plain terms, it means that each of the 12 standard keypad buttons produce a very distinct two-tone output. Each button on a tone dial produces a dual-tone output whose frequencies are determined by the row and column in which the button is located. If you require to refresh your memory or need a better understanding of the DTMF concept, please flip back a few pages to chapter 1 where detailed explanations and illustrations of tone dials are presented. To grasp the operation of this rather complex design, a thorough understanding of a tone dial and DTMF schemes is a must, or the following explanation will be just a blur in your mind.

THE DTMF RECEIVER IC

The M-957-01 DTMF Receiver (see Fig. 14-2) is manufactured by Teltone Inc. (P.O. Box 657, 10801-120th Avenue Northeast, Kirkland, Washington 98033). Like the familiar G8870, the M-957 receiver detects and converts the standard DTMF tone into the binary equivalent. The M-957 incorporates all the needed filtering within its 22-pin plastic body. It determines if the input signal is just noise, a speech pattern, or a dialing signal that is within a predetermined ± 2 percent of the required DTMF bandwidth of 697 to 1633 Hz.

The 957 processes the conventional DTMF Touch-Tone frequencies and provides a four-bit binary output, and a strobe (see Table 14-1) output that goes high when a valid DTMF code is received. The strobe signal is used by the Tele-Guard II when converting the binary signal to a decimal output.

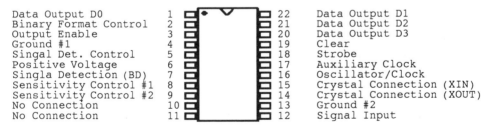

Data Output D0	1		22	Data Output D1
Binary Format Control	2		21	Data Output D2
Output Enable	3		20	Data Output D3
Ground #1	4		19	Clear
Singal Det. Control	5		18	Strobe
Positive Voltage	6		17	Auxiliary Clock
Singla Detection (BD)	7		16	Oscillator/Clock
Sensitivity Control #1	8		15	Crystal Connection (XIN)
Sensitivity Control #2	9		14	Crystal Connection (XOUT)
No Connection	10		13	Ground #2
No Connection	11		12	Signal Input

Fig. 14-2 *The pinout assignments for the Teltone M-957 DTMF Receiver IC.*

Table 14-1. An extensive listing of pin functions for the M-957 IC.

Pin Functions of the M-957 DTMF Receiver

Pin	Function
Signal In	DTMF input. Internally bias so that the input signal may be AC coupled. SIGNAL IN also permits DC coupling as long as the input doesn't exceed the positive supply.
12/$\overline{16}$	DTMF signal detection control. When 12/$\overline{16}$ is at logic "1", the M-959 detects the 12 commonly used DTMF signals. When 12/$\overline{16}$ is at logic "0", the IC detects all 16 DTMF signals.
A,B	Binary DTMF signal sensitivity control inputs. A and B select the sensitivity of the SIGNAL IN input to a maximum of -31dBm.
D0,D1,D2,D3	Data output. When enabled by the OE input, the data outputs provide the code corresponding to the detected digit in the format programmed by the HEX pin. The data outputs become valid after the a tone pair has been detected and are cleared when a valid pause is timed.
OE	Output Enable. When the OE is at logic "1", the data outputs are in the CMOS push/pull state and represent the contents of the output register. When OE is at logic "0", the data outputs are forced to the high impedance ot the third state.
HEX	Valid data indication. STROBE goes to logic "1" after a valid tone pair is sensed and decoded at the data outputs. STROBE remains at logic "1" until a valid pause or a CLEAR input is driven to logic "1", which ever is earlier.
CLEAR	STROBE Control. Driving CLEAR to logic "1" forces the STROBE output to logic "0". When CLEAR is at logic "0", STROBE is forced to logic "0" only when a valid pause is detected. Tie to VNA or VND when not is use.
BD	Early signal presence output. BD indicates that a possible signal has been detected and is being validated. BD precedes STROBE and the data outputs.
XIN, XOUT	Crystal connections. When an auxiliary clock is used, XIN should be tied to logic "1".
OSC/$\overline{CLK}$	Time Base Control. When OSC/$\overline{CLK}$ is at logic "1", the output of the IC's internal oscillator is selected as the time base. When OSC/$\overline{CLK}$ is at logic "0" and the XIN is at Logic "1", the AUXCLK input is selected as the time base.
AUXCLK	Auxiliary Clock Input. When OSC/$\overline{CLK}$ is at logic "0" and the XIN is at logic "1", the AUXCLK input is selected at the IC's time base. The auxiliary input must be 3.58MHz, divided by 8 for the IC to operate to specifications. If unused, AUXCLK should be left open.
VNA, VND	Negative analog and digital power supply connections.
VP	Positive power supply connections.
N/C	No Connection. These pins have no internal connection and may be left floating.

HOW IT WORKS

Figure 14-3 is a block diagram of the complete Tele-Guard security device. Notice that the telephone is normally powered by the talk battery, which is actually a + 12 V power source within the Tele-Guard. Entering the correct code on the phone keypad eventually causes relay RY1 to pull in. Relay poles P1 and P2, which are part of RY1, switches the telephone through the off-hook detector directly to the central office telephone line, where a dial tone is received. At this time, normal telephone dialing can take place. As you will see shortly, opto-couplers ensure that there can be no direct connection between the Tele-Guard power supply and the telephone line.

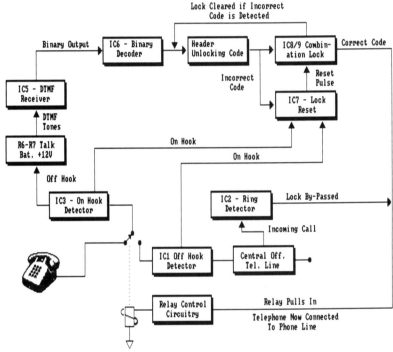

Fig. 14-3   *A block diagram illustrating how the digital telephone lock operates.*

The talk battery 12 V and the 6 V source for the Tele-Guard circuits are provided by the power supply shown in Fig. 14-4. The voltage regulators (IC10 and IC11) should be of the TO-3 diamond configuration (see Fig. 14-5). TO-3s are capable of handling more current and should be used for both the talk battery and the main 6 V power supply. Heatsinks are also recommended.

As shown in Fig. 14-1, the Tele-Guard internal combination lock is in a reset state when the device is in its normal standby condition. When the telephone handset is lifted from its cradle, the internal transistor of the opto-isolator (see Fig. 14-6) off-hook Detector (IC3) conducts and brings pin 5 of that IC low. The low is applied to inverter IC4-a (pin 1), which changes the low to a

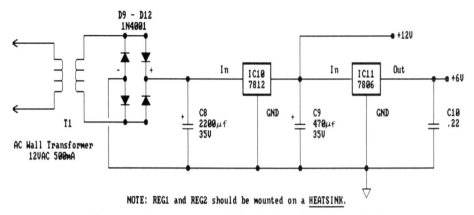

Fig. 14-4   *The power supply for the digital telephone lock.*
You might want to use TO-3 style (diamond shape) for both voltage regulators.

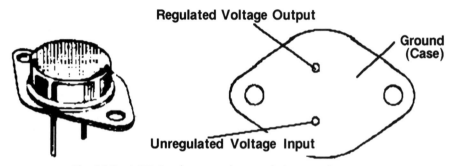

Fig. 14-5   *A TO-3 voltage regulator and pinout assignments.*

Internal LED Anode	1		6	Transistor "Base"
Internal LED Cathode	2		5	Transistor "Collector"
No Connection	3		4	Transistor "Emitter"

Fig. 14-6   *Pinout assignments for the 4N33 opto-isolator.*

high. The high is applied to switching diode D3, which blocks the high. Therefore, pins 4 and 10 of both IC8 and IC9 (D-type flip/flop see Fig. 14-7) are floating. The floating pins allow the ICs to change state from high to low when a valid unlocking code is present. More on those two IC's a little later.

For the proper operation of the Tele-Guard, you must apply a talk battery. In this instance, a talk battery is a pure + 12 V signal that is applied to the telephone that is to be secured. (The telephone is connected to the telephone lock via a modular socket SO1.) The talk battery is applied to SO1 through opto-isolator (4N33) IC3 and resistor R6 (120 Ω, 1W).

The telephone red and green wires, located within the phone line cord, are connected through contacts P1 and P2, which are part of relay RY1. The resting position of P1 connects the telephone to the talk battery through resistor R7. The electrical path to ground is provided by P2 and by resistor R6.

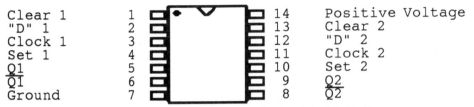

Clear 1	1	14	Positive Voltage
"D" 1	2	13	Clear 2
Clock 1	3	12	"D" 2
Set 1	4	11	Clock 2
Q1	5	10	Set 2
$\overline{Q1}$	6	9	Q2
Ground	7	8	$\overline{Q2}$

Fig. 14-7 *Pinout assignments for the 7474 dual D flip/flop.*

MAKING A CALL

To show how the system works, assume that the number 8 button is being pressed, thereby applying an 852/1336 Hz DTMF signal to pin 12 of IC5, the DTMF receiver.

As seen with other projects that require the G8870, the M-957 also converts this signal into its binary equivalent (1000). In addition to the binary output, IC5 also produces a strobe high for as long as the number 8 keypad button is pressed. (Pin 18 goes low when the button is released.)

Both the binary and strobe outputs are applied to IC6, a 4 to 16 decoder/ demultiplexer (see Fig. 14-8). With the DTMF binary output applied to pins 20 and 23 of IC6, and pin 19 held low, the decimal equivalent of the input will appear at IC6 output pins (pins 2 to 11). In this instance, pin 9 of IC6, which corresponds to an input binary code for the number 8, will go low. All other output pins remain high.

Inverter IC4-b (see Fig. 14-9) is used to invert the high strobe coming from IC5 into a low strobe for IC6 because pin 9 of IC6 requires a low in order to pass the binary equivalent of its input.

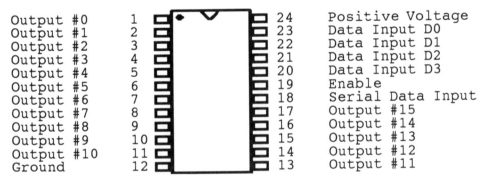

Output #0	1	24	Positive Voltage
Output #1	2	23	Data Input D0
Output #2	3	22	Data Input D1
Output #3	4	21	Data Input D2
Output #4	5	20	Data Input D3
Output #5	6	19	Enable
Output #6	7	18	Serial Data Input
Output #7	8	17	Output #15
Output #8	9	16	Output #14
Output #9	10	15	Output #13
Output #10	11	14	Output #12
Ground	12	13	Output #11

Fig. 14-8 *Pinout assignments for the 74154 (1-of-16 data distributor IC).*

Inverter #1 Input	1	14	Positive Supply Voltage
Inverter #1 Output	2	13	Inverter #6 Input
Inverter #2 Input	3	12	Inverter #6 Output
Inverter #2 Output	4	11	Inverter #5 Input
Inverter #3 Input	5	10	Inverter #5 Output
Inverter #3 Output	6	9	Inverter #4 Input
Ground	7	8	Inverter #4 Output

Fig. 14-9 *Pinout assignments for the 7404 hex inverter.*

THE UNLOCKING CODE

The section that actually senses the unlocking code is SO2, a 20 pin header socket that is jumped to correspond to the unlocking code. The programming of this header is covered in this chapter. But for now, let's continue.

The unlocking sequence is determined by IC8 and IC9, two dual D-type edge-triggered flip/flops wired as a sequential pass on. The sequential pass-on is a series of four independently clocked, cascaded D-type flip/flops. The unlocking pulse, which is available at pin 9 of IC9, will go low—thereby causing RY1 to switch the secured telephone to the telephone line—only if the unlocking digit 1 is followed by digit 2, then by digit 3, and finally by digit 4. Each unlocking digit will provide a low from IC6 through SO2 to its corresponding pin on IC8 and IC9.

A low must be applied to pins 4 and 10 of both IC8 and IC9 in order for both to change states. The low can be applied from on-hook detector IC3, or from IC7, which functions as a mis-dialed digit reset.

Pin 9 of IC9 goes low when the correct four digit unlocking code is entered. The low is applied to the base of Q1 through diode D7, turning both transistor Q1 and LED4 on, thereby energizing relay RY1.

Contact P1, which is controlled by RY1, connects the telephone set to the telephone line through modular plug PL1, thereby providing dial tone and normal operation. Also, when P1 closes it opens pin 2 of IC3, which forces the output of IC4-a low. That in turn forces pins 4 and 10 of IC8 and IC9 low, which resets their flip/flops to the standby condition.

When the telephone connects to the line it also energizes on-hook detector IC1, causing pin 5 of IC1 to go low. The low is applied to Q1 through diode D5, which keeps RY1 energized during a normal telephone conversation.

MAINTAINING CONNECTION

When using a telephone, you might have noticed the sound of clicks in the receiver when the central office connects you to the called party, especially if the equipment uses antiquated relay switching. The clicks are an interruption of the voltage to the telephone instrument. If no means of counteracting the momentary voltage loss were applied to our circuit, the small internal LED across pins 1 and 2 of IC1 would go out and the IC1 output would reverse, causing RY1 to drop out and terminating the call in progress.

Capacitor C7, which is connected between the base of Q1 and the +6 V supply, prevents RY1 from releasing. The capacitor charges when a ground (low) is available at the Q1 base. During a brief interruption in the telephone line voltage, the capacitor discharges and applies a low (ground) to the Q1 base, thereby keeping RY1 energized during the entire interruption.

When the telephone call is complete, returning the handset to the cradle disconnects the telephone line from the on-hook detector, thereby restoring IC1 pin 5 to a high, which cuts off Q1, causing RY1 to drop out. This drop-out returns the telephone to its standby condition.

INCOMING CALLS

The security code is not needed to answer an incoming call because Tele-Guard II has an override circuit that allows the secured telephone to answer a ringing signal.

The override circuit is based on opto-isolator IC2, which functions as a ring detector. A telephone ringing signal, which appears across the red and green telephone wires, is approximately 90 to 100 V at a frequency between 20 and 30 Hz. If you trace the circuit shown in Fig. 14-1, you will find that a series circuit consisting of capacitor C4, resistor R5, and IC2, is connected across the telephone line. When the ringing signal is received, the IC2 internal LED begins to flash, which causes pin 5 of IC2 to go low. That low is then applied through diode D6 to the base of Q1, which causes this transistor to conduct and energize the RY1 relay. When the relay pulls in, the contacts then connect the telephone to the line so that the call can be answered.

Unfortunately, there is a problem with this arrangement. Pin 5 of IC2 goes low only when its internal LED flashes, which means that the telephone can only be answered during the ringing half of the signalling cycle. To overcome this dilemma, R4 and C3 are added to provide a steady low to the base of Q1 whenever pin 5 of IC2 attempts to go high. The R4/C3 time constant is approximately four seconds, which is more than enough time to compensate the off period of the ring signal.

WIRING THE HEADER (SO2)

The wiring of SO2 determines the specific digits and their sequence needed to unlock the telephone. For simplicity, the required SO2 connections have been broken down into the wiring groups shown in Fig. 14-10.

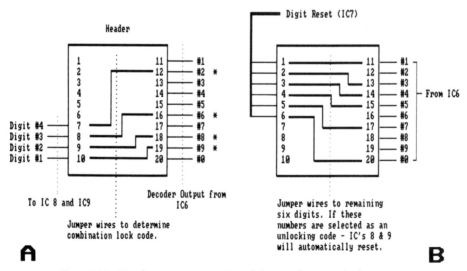

Fig. 14-10 For the proper operation of the combination lock circuitry, a 20-pin header must be wired as shown. You can change the combination if you wish.

Figure 14-10 is a top view of the SO2 20-pin header assembly. Pins 1 to 6 are used for the mis-dialed digit reset pulse (more on that a little later). Pins 7 to 10 are the selected unlocking digits, where pin 10 is the first digit that is entered, pin 9 is the second, pin 8 is the third, and pin 7 is the fourth and last number.

Pins 2 to 11 are connected directly to the decoded decimal output from IC6 where, in the standby mode, they are all high. Before any wiring can take place, you must determine the four-digit combination. Any digit from 0 to 9 can be used. The combination can contain double digits, such as 2234, 4499, and so on.

For example, assume you wish to use the combination:

9862

The combination can also be written as shown in Table 14-2.

Table 14-2. An easier way to visualize how
to wire the digital combination for the telephone lock.

Header Connections		
Digit Place	Combination	Header Pin Number
Digit #1	9	19
Digit #2	8	18
Digit #3	6	16
Digit #4	2	12

As shown in Fig. 14-10, using jumper wires, connect header pin 19 to pin 10. That will indicate to the Tele-Guard that number 9 will be the first digit of the unlocking sequence.

Connect header pin 18 to pin 9. This wiring will indicate that the number 8 is to be the second unlocking digit.

Connect header pin 16 to pin 8. That designates that the third digit is to be the number 6.

And finally, connect pin 12 to pin 7. This indicates that the number 2 is the fourth digit of the selected combination.

The remaining SO2 wiring is the additional wiring needed for the digit-reset pulse. Connect the remaining six unused digits (pins 11, 13, 14, 15, 17, and 20) to header pins 1 through 6.

When the SO2 wiring is completed and checked for accuracy, snap on the header cover and write the combination on the top. You might want to consider building a number of header assemblies so that the combination can be varied from time to time.

THE RESET PULSE

As indicated above, the Tele-Guard II can provide maximum security to your telephone by making use of a circuit that will reset the D-type flip/flop (IC8 and IC9) every time an incorrect digit is entered as part of the unlocking sequence.

The heart of the reset circuit is IC7, an 8-input NAND gate (see Fig. 14-11). To refresh your memory, the output pin of an NAND gate will go high when any of its eight inputs goes low.

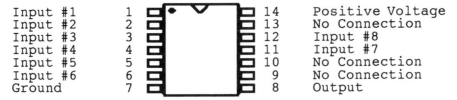

Input #1	1		14	Positive Voltage
Input #2	2		13	No Connection
Input #3	3		12	Input #8
Input #4	4		11	Input #7
Input #5	5		10	No Connection
Input #6	6		9	No Connection
Ground	7		8	Output

Note: Diagram can also be used for the 74LS13 Schmitt Trigger NAND Gate

Fig. 14-11 *The pinout assignment for the 74LS30 8-input NAND gate.*

For Tele-Guard II, header pins 1 to 6 are connected directly to IC7. The additional two IC7 input pins, which are not used, are connected to the + 6 V power supply. In that state, all inputs are high so the output is low.

For example, assume that the button number 9 is pressed on a telephone keypad. That causes header pin 19 to go low as long as the button is depressed. The jumper wire from header pin 19 passes the low to header pin 10, which is connected to the sequential pass on circuitry (IC8 and IC9). At this point, all is operation well.

Assume that the second digit entered via the keypad is the number 3. By referring to Fig. 14-1, you can see that header pin 13 is pulled low. The jumper from pin 13 also pulls pin 2 low. The number 3 digit is considered to be a mis-dialed entry due to the fact that our code requires the second number to be entered will be an 8, not a 3 or any other number.

Because the number 3 puts a low on header pin 2 and because pin 2 is con-nected to one of the NAND gate IC8 inputs, IC output (pin 8) goes high. This high is inverted by IC4 to a low, which is passed on to IC8 and IC9 via pins 4 and 10. Both IC8 and IC9 reset when pins 4 and 10 are low, thereby cancelling the previously saved unlocking digit information. To bypass the Tele-Guard, the user must enter the complete four digit code in the proper sequence.

VISUAL INDICATOR

LEDs 1 to 4, shown in Fig. 14-1, provide a visual indication of the unlock-ing sequence. They light in sequence as the correct combination is entered. When the fourth digit is entered, LEDs 1 to 3 extinguish and LED4 goes on, indicating that the telephone has been granted access to the outside line.

CONSTRUCTION

The Tele-Guard II is assembled on a double-sided PC board; both templates (one for the solder side and the second for the component side), are shown in Figs. 14-12 and 14-13 respectively. Alternatively, an etched and drilled boards can be purchased from the source given in the parts list for only $26.00 (shipping included). Take note that the PC board, whether homebrewed or purchased, does not have plated-through holes. After a component is installed, you must solder the component lead to the traces on both sides of the board (if present).

Using Fig. 14-14 as a reference, install the resistors and the capacitors first. Then the solid state devices and the relay. Use sockets IC5 and IC6 and for the header SO2, because you wish to change the combination periodically. Make certain that heat sinks having at least 2-1/4 square inches in area are installed on voltage regulators IC10 and IC11. They will generate quite a large amount of heat.

To conserve precious PC board real estate, diodes D9 to D12 must be mounted on end. Just make sure that their polarity is correct before soldering.

The complete unit can be installed in just about any kind of cabinet that you might have. For that professional look, I have included in Fig. 14-15 a template for the front panel of the housing. Just make a copy of this template on a copy machine, trim to size, and fasten it to the front.

CHECKOUT

When all components are installed, soldered (on both sides if needed), and checked for proper polarity, plug in the ac wall transformer T1 (12 to 18 Vac, 500 mA) and check for proper voltages at the power supply.

If everything checks out to your satisfaction, connect the telephone to be secured to terminal SO1. Then, using the keypad, key in the programmed security code and note the lighting of LEDs 1 to 3.

When the fourth digit is entered, you should hear the relay pull in. Simultaneously, LED4 should light while LEDs 1 to 3 go out. (The relay will only pull in and then drop out at this point.)

Connect the red and green wires of a modular telephone cord to their proper locations on the PC board and plug the other end into the normal telephone modular wall connector (See Fig. 14-16 for suggested one phone installation).

Again, using the proper unlocking code, enter the four security digits. When the fourth digit is entered, the RY1 relay will pull in and this time stay in. At this time, you will hear the dial tone in the receiver.

To further test out the operation of the project, hang up the telephone; then lift the handset again. This time enter the first two digits of the unlocking code. Notice that LED1 and LED2 will light up. Then press any other button that is not associated with the code. This time both LEDs should go out.

To test the on-hook circuit, lift the handset and enter one or more of the

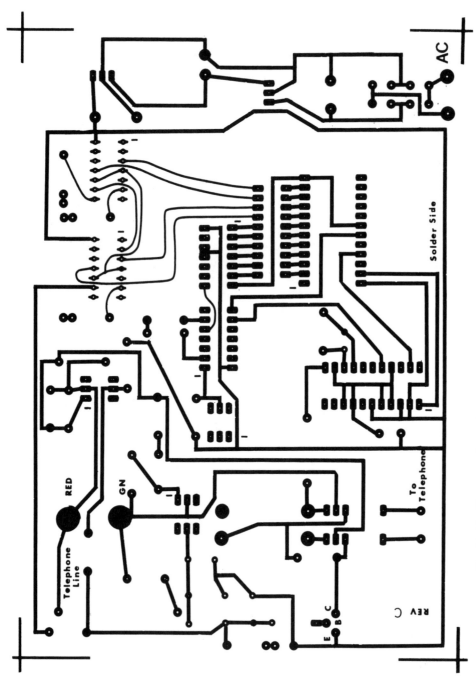

Fig. 14-12 *The artwork needed to make the solder side of the telephone lock PC board.*

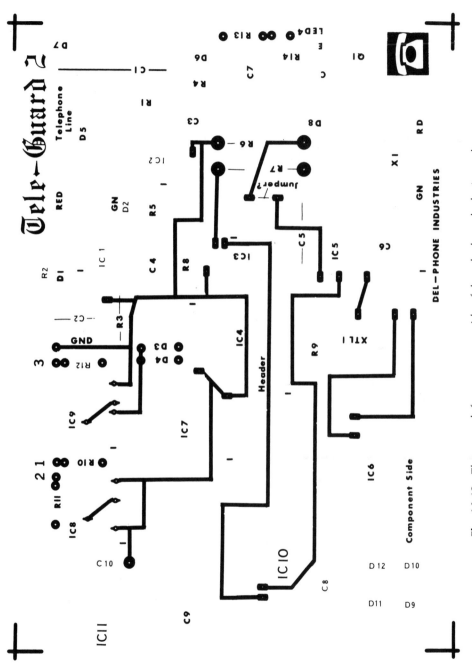

Fig. 14-13   The artwork for component side of the telephone lock. Artwork for both the solder and component sides of the PC board are shown here at a scale of 1 to 1.

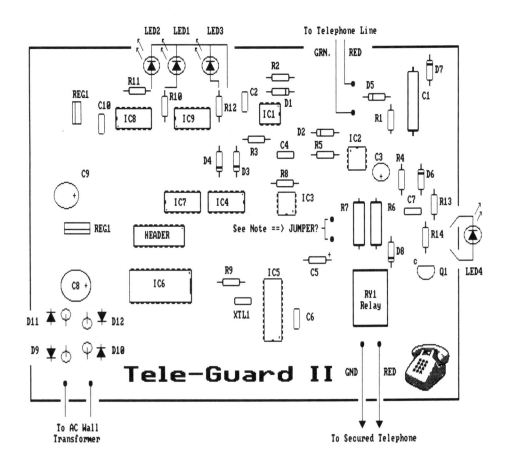

NOTE: ⟨JUMPER?⟩ should only be connected if Tele-Guard II is to use the
      Dial Pulse Counter Circuit. Otherwise leave OPEN.

Fig. 14-14   *The component layout for the digital telephone lock.*
The circuit uses a large number of parts. Be careful when stuffing the board.

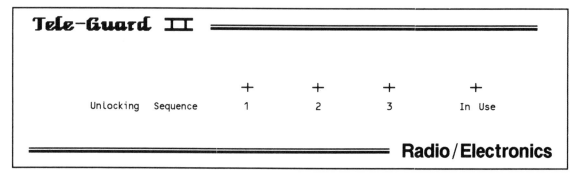

Fig. 14-15   *The front panel template for the digital telephone lock.*

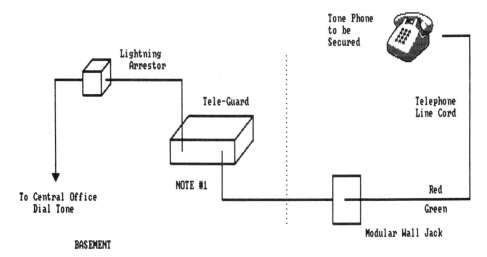

Tone Phone
to be
Secured

Lightning
Arrestor

Tele-Guard

Telephone
Line Cord

NOTE #1

To Central Office
Dial Tone

Red

Green

Modular Wall Jack

BASEMENT

NOTE #1: Break Telephone Line between Lightning
arrestor and Modular Wall Jack. Insert
Tele-Guard II at this point as shown.

Fig. 14-16 *Connecting the digital lock to a single-line telephone.*

digits needed to unlock your telephone; then hang up. The LEDs that have been turned on will extinguish, indicating that the counters were reset.

To test the security override circuitry, have a friend call you. When the telephone rings, note that LED4 will light during the ringing half of the signalling cycle. Then lift the handset; you should now be able to carry on a conversation without any further action on your part.

MULTITELEPHONE SYSTEM INSTALLATION

Figure 14-17 shows a multitelephone home installation. Notice that all telephones in the home have been connected in parallel. Make certain you observe the red-green polarity but keep in mind that telephone installers have been known to make mistakes that are never corrected. So you might find that the red and green wires have been exchanged somewhere between the phone jack and the central office. If you are in doubt about the connection, check it out by using a telephone as your test set.

PULSE-DIAL RETROFIT ASSEMBLY

Although the Tele-Guard was specifically designed for use with a Touch Tone telephone, it can also provide the same kind of security for rotary and push-button, pulse-dialed telephones. This list also includes the now widely used cordless telephone. Because there are so many telephones of this type in use today, the security of the cordless instrument is much to be desired. By incorporating the Tele-Guard security device with your cordless telephone, the

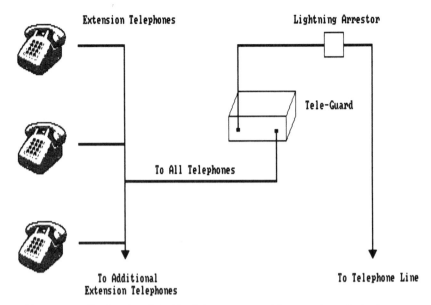

Fig. 14-17   *Sharing the digital lock in a system with a number of telephones.*

user must enter the security code before any dialing can begin. This will prevent a neighbor (who also has a cordless telephone) from accessing your line and making calls using your line, which, of course, you will be billed for. Just by installing this small dial pulse counter and interfacing it with the mother board, high security can be obtained with all types of rotary as well as tone-dial telephones.

The retrofit consists of a few easy-to-do modifications to the original Tele-Guard II project. One of the modifications is a small subassembly, shown in Fig. 14-18 that is used as a substitute for the DTMF receiver IC (IC5). To avoid confusion, the retrofit uses the part number sequence from the original Tele-Guard; hence, the subassembly integrated circuits are labelled IC12 and IC13.

PULSED DIALING

The heart of the subassembly is IC12, Teltone's M-959 dial-pulse counter (see Fig. 14-19), which converts voltage pulses corresponding to rotary phone dialing digits into a binary coded output. It is essentially a direct substitute for the original DTMF receiver (IC5), except, instead of voltage pulses, the DTMF receiver converts audio tones that represent Touch Tone digits into a binary coded output.

IC13 is the trigger for IC12. It monitors the talk battery line for dial-pulse signalling and hook status. When the telephone is on-hook, the IC13 output is an active low. When the telephone goes off-hook, the output goes high, thereby triggering IC12 on/off hook and operating timers, which, in turn, trigger the IC12s internal counter, digit decoder and output control.

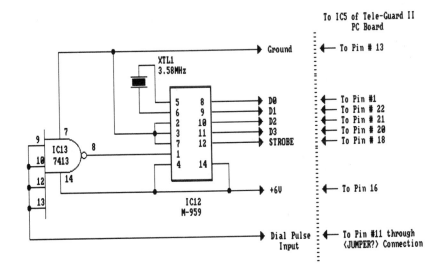

To IC5 of Tele-Guard II
PC Board

←— To Pin # 13

←— To Pin #1
←— To Pin # 22
←— To Pin # 21
←— To Pin # 20
←— To Pin # 18

←— To Pin 16

←— To Pin #11 through
⟨JUMPER?⟩ Connection

If Dial Pulse Counter is to be used
with Tele-Guard II circuit, connect
these points to Tele-Guard's IC5
location.

NOTE: If Dial Pulse Counter is to be used with Tele-Guard II
solder a solid bus wire on Tele-Guard's PC Board at
===> JUMPER? <=== location.

Fig. 14-18 *Converting your digital telephone lock to operate in a pulse line.*

L̄C̄	1	14	Positive Supply Voltage
Clear	2	13	Off Hook
10/20	3	12	Strobe
OE	4	11	Data 3 Outout
Xout	5	10	Data 2 Output
Xin	6	9	Data 1 Output
Ground	7	8	Data 0 Output

Fig. 14-19 *The pinout assignments for the M-959 dial-pulse counter.*

PRINTED CIRCUIT

The subassembly is built on a small printed circuit board. A full size template is presented in Fig. 14-20.

The parts layout is shown in Fig. 14-21. Note, in particular, that the two ICs are mounted opposite; that is, pin 1 of both ICs point in the opposite directions.

The double row of holes on the subassembly PC board corresponds to the IC5s pin pattern on the original Tele-Guard II mother board. Make certain that none of the holes get filled with solder during the assembly. (Obviously, IC5 is not needed if you are building the entire pulse-dialing telephone lock from scratch.)

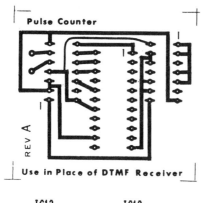

Pulse Counter

REV A

Use in Place of DTMF Receiver

Fig. 14-20 PC board artwork for the dial-pulse counter (scale is 1 to 1).

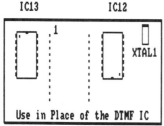

IC13   IC12

1

XTAL1

Use in Place of the DTMF IC

Fig. 14-21 Parts layout for the dial-pulse counter.

Complete the assembly except for the 3.5 MHz crystal, XTAL1. The crystal is the same one used in the DTMF Tele-Guard. If you are doing a retrofit, the crystal must be removed from the mother board. If you are building from scratch, simply install XTAL1 directly on the subassembly.

For now, set the subassembly aside until you complete a few modifications to the main Tele-Guard II PC board.

MODIFICATIONS TO THE MOTHER BOARD

The modifications are shown in Fig. 14-22 with dotted lines around them. You can enter the changes on the original schematic, or photocopy Fig. 14-1 and paste it over the original full schematic.

From the mother board, remove XTAL1, IC5, C5, and R9. If you remove the solder from each connection using one of the available desoldering braids, the parts will literally fall off into your hand—without damage to either the part or the printed circuit trace. The best and the easiest solder removal will be attained using a braid about 1/8-inch wide. Double check to be sure that every hole at IC5 location is open. Install the crystal on the subassembly PC board.

A pulse dial actually connects and disconnects the telephone set from the line in rapid succession. If the make-break sequence was not compensated, the very first pulse would cause the Tele-Guard to disconnect the telephone from the line. To prevent an automatic disconnect, tack solder at 330 ΩF, 16 V radial lead electrolytic capacitor across resistor R8. (In keeping with the policy of using the part number sequence from the original project, the capacitor is C11.)

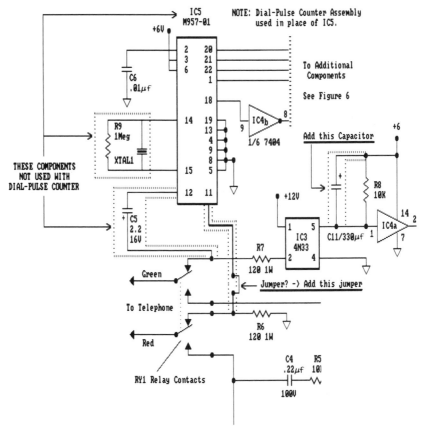

Fig. 14-22 *To incorporate the dial-pulse
counter in the telephone lock, a few modifications must be made.*

If there is sufficient clearance between the mother board and its metal cabinet, you can install C11 on the foil side of the board.

The polarity of C11 must be observed. When installed across R8, the C11s positive lead should point toward IC4; the negative lead should point toward IC2.

Install a jumper on the main PC board across the two empty solder pads that are located directly adjacent to R7 (on the side opposite R6). The jumper connects the ungrounded side of R6 to pin 11 of IC5, which is used only as an interconnection to the small subassembly.

Use Fig. 14-23 for reference and solder a 22 solid, 1-1/2 inch long, uninsulated wires in IC5 holes. (They will be cut to length after the subassembly is installed.) The wire size can be number 20, 22, or 24. Note, from Fig. 14-23, that some of the wires make connection to traces on both sides (the top and bottom) of the main board, so make certain that you solder both sides (the holes are not plated through). After soldering the wires, bend them upwards, at a right angle to the board.

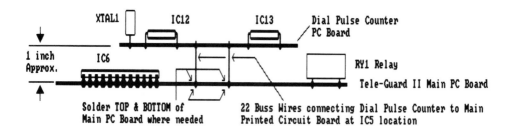

Dial Pulse Counter Installation

Fig. 14-23   *Installing the dial-pulse counter in place of the DTMF IC.*

Again, using Fig. 14-23 as a reference, position the subassembly directly over the 22 wires and slide the subassembly on the wires. Make certain that each wire passes through its corresponding hole. Position the subassembly so that it does not touch anything on the main board and then solder the wires to the subassembly solder pads. That will complete the retrofit (Note: If it is more convenient, install the wires on the subassembly first. Then pass them through the holes in the main board.)

Instead of using individual wires to connect the subassembly to the main board, you might want to use a wire-wrap IC socket. Make sure you leave enough room for soldering.

PROGRAM AND CHECKOUT

Program the main PC board with the security code for the pulse dial Tele-Guard II exactly as it was described for the Touch Tone version. Everything else remains the same. You gain access to the telephone line by first pulse-dialing the four-digit security code. Within a second or two a dial tone will be heard from the receiver; you then can dial the desired telephone number.

Using the digital telephone lock with cordless telephones is the same as using it with a cord telephone. Turn on the wireless receiver and dial the security code. Once accepted, a dial tone will be heard in the handset. Normal cordless dialing and conversation can then take place without worrying about neighbors making free long-distance calls on your line.

PARTS LIST FOR THE DIGITAL TELEPHONE LOCK

R1, R14	1000Ω resistors
R2, R10 – R13	220Ω resistors
R3	47kΩ resistors
R4	220kΩ resistors
R5, R8	10kΩ resistors
R6, R7	120 Ω 1 W carbon resistor
R9	1 MΩ resistor

C1	2.2 μF, 250 V, nonpolar capacitor
C2, C4	22 μF, 100 V electrolytic capacitors
C3	1000 μF, 10 V radial electrolytic capacitor
C5	2.2 μF, 16 V axial electrolytic capacitor
C6	0.01 μF, 100 V disc capacitor
C7	47 μF, 16 V radial electrolytic capacitor
C8	2200 μF, 35 V radial electrolytic capacitor
C9	470 μF, 35 V radial electrolytic capacitor
C10	0.22 μF, 50 V disc capacitor
IC1 – IC3	4N33 opto-isolator
IC4	7404 hex inverter
IC5	M-957-01 DTMF receiver (Del-Phone Industries)
IC6	74154 4-to-16 decoder
IC7	74LS30, 8-input NAND gate
IC8, IC9	7474 dual, D-type flip/flop
IC10	LM7812, 12 V voltage regulator (TO-3 case)
IC11	LM7806, 6 V regulator (TO-3 case)
Q1	2N4402 transistor
D1, D2,	
D8 – D12	1N4001 diodes
D3 – D7	1N914 diodes
LED1 – LED4	Light-emitting diodes
PL1	Modular plug with attached cord (Radio Shack)
RY1	6 V dc DPDT (double-pole, double throw) relay
SO1	Modular telephone socket (Radio Shack)
SO2	20-pin header with cover (Jameco Electronics)
T1	Wall transformer 12-18 V ac 500 mA
XTAL1	3.58 MHz crystal
1	Printed circuit board (Del-Phone Industries)
1	Telephone line cord (modular type—Radio Shack)
1	Housing
2	Heatsinks for voltage regulators
1 or more	Touch Tone telephones to be secured

PARTS LIST FOR DIAL-PULSE COUNTER (RETROFIT ASSEMBLY)

C11	330 μF 16 V radial electrolytic capacitor
IC12	M-959 dial-pulse counter (Del-Phone Industries)
IC13	74LS13 Schmitt trigger IC
XTAL1	3.58 MHz crystal
1	Printed circuit board (Del-Phone Industries)

☎15
The Talking
Telephone Ringer

BELLS HAVE BEEN USED AS SIGNALLING DEVICES FOR TELEPHONES PRACTICALLY EVER since their invention over a century ago. Through the years, little has changed in the basic construction and operation of this annoying and irritating device. There was little change until the 1980s, when a landmark court decision opened the doors that allowed billions of dollars of inexpensive telephone items to flood the U.S. markets.

Well, that is progress. It is progress that has not only given U.S. consumers cheap imports, but it also ushered in the era of throw-away telephones. To most Americans, this might seem to be the end of U.S. involvement in the telecommunications industry. But to the contrary, this court decision planted the seeds for additional technological advancements, not only in the telcom arena, but in every part of electronics. One very good example can be seen in chapter 1, *Telephone Basics*. A standard phone consists of a dial, a ringer, a network, and a handset. Foreign as well as U.S. manufacturers have designed integrated circuits that replace the entire internal electronics of a telephone. These ICs, called LSI (large scale integration), have all the electronics needed to make a telephone squeezed into an area no larger than your thumb nail. Yes, this also includes a ringer circuit.

With the technology available today, engineers have gone wild designing telephones in all shapes and sizes. Clanging bells have been replaced with the soothing resonance of electronically generated sounds. If you wish, you can purchase a telephone that quacks like a duck.

With the author's own design (animated telephone ringer), you can have a toy dog bark and wag its tail to indicate an incoming call.

Hey folks! This is the decade of the 90s. The twenty-first century is almost here. So how about taking some of the technology in this book and doing something really wild with it? Instead of having a telephone ring generate a wabble tone. Or have a duck quack or a dog bark. Make your telephone talk to you.

A talking telephone? It might seem far fetched, but it can be done. In fact you can do it very easily with a handful of parts. By taking an IC called a Digi-talker, you can have this speech synthesizer say any word you wish when an incoming ring signal is detected (of course the word must be included in the

Digitalker vocabulary). To further enhance the project, add a two-bit counting circuit and a programmed 74188 IC. The result will be a telephone ringer that will say four different words each and every time an incoming ring is generated by the central office. Sounds great, right? In the next chapter, you can see what else can be done with this truly remarkable integrated circuit.

HOW THE PROJECT WORKS

Figure 15-1 shows the basic wiring of the speech synthesizer circuit. The heart of this circuit is the Digitalker IC (SPC MM54104). This device, manufactured by National Semiconductor, can take a binary input that corresponds to a preprogrammed word and have it, using its internal electronics, convert this raw data into a highly intelligible voice. If the Digitalker is the heart of the circuit, the MM52164 ROM (read-only memory) is the brain. Here, compressed digital data is stored and is instantly available whenever the Digitalker requires it.

Other ICs needed to create this circuit are the LM346 and the LM386. The resistors and capacitors associated with the LM346 (IC2) are used as a filter to further enhance the quality of the speech pattern. The LM386 (IC3) is a standard audio amplifier that transforms the minute audio input into a room-filling synthesized output.

To make the construction of the talking telephone ringer as painless as possible, Fig. 15-1 is the main speech circuit for both projects in this chapter.

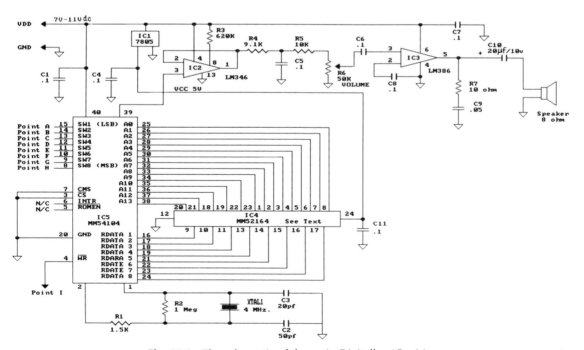

Fig. 15-1   *The schematic of the main Digitalker IC wiring.*

All additional interface electronics will be connected to point A through point I (pins 8 to 15 and pin 4). Points A to H is for the binary data input. It is here that you input the proper code for each individual word you wish your circuit to say. Point I (pin 4) is the start-up pin (write). If you apply a legal code at the input of the Digitalker and place a positive-going pulse on pin 4, the synthesizer will say the word you have just programmed at its input. Starting to sound a little bit easier?

Figure 15-2 is your first telephone interface. Here, the ringing signal from the telephone line is placed across resistors R1 and R2. It is rectified by the bridge circuit (D1 to D4); then it is delivered to the internal LED of IC1 (4N37).

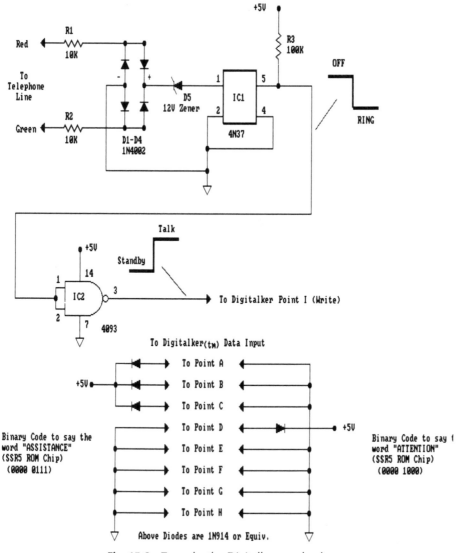

Fig. 15-2 *To make the Digitalker speak when an incoming call is being received, an interface circuit must be constructed.*

When no ring voltage is present, the output of IC1 (pin 5) will be about 5 V. But when a ring is detected, this voltage drops to a logic 0.

Because you would like your circuit to provide a snap-action output pulse for the Digitalker, IC2 is used. The 4093 will not only provide a Schmitt trigger output, it will also provide an inverted signal as well (no ring—logic 0 output; detected ring—logic 1 output). By taking the output of your NAND gate (IC2 pin 3), you can control the start-up function of the speech synthesizer. Every time a ring signal is present, IC2 will provide a snap-on, positive-going pulse. This pulse is then applied to pin 4 (point I) of the Digitalker.

So far you have a means of governing the speech output with your interface. But what about programming a word into the chip? At the bottom of Fig. 15-2 you can see a number of diodes and ground connections. If you take a + 5 V signal and deliver it to the Digitalker via a number of glass diodes (points A to C), these pins will have a logic 1 level applied to them. Now by taking the remaining data inputs and grounding them (points D to H), these pins can be considered to be a logic 0. What you have is:

	MSB							LSB
Points	H	G	F	E	D	C	B	A
	0	0	0	0	0	1	1	1

Where MSB is the most significant bit and LSB is the least significant bit.

This code corresponds to the word *assistance* (using the SSR5 memory chip. More on this a little later). With the code 00000111 applied to the data input of the Digitalker and by having a start-up pulse applied to pin 4, you now have the circuit say the word *assistance* every time the output from IC2 (pin 3) goes from ground to a positive 5 volts.

To make your talking circuit say the word *attention*, all you have to do is to apply + 5 V to point D while all other points are grounded. At this time, you have:

	MSB							LSB
Points	H	G	F	E	D	C	B	A
	0	0	0	0	1	0	0	0

This code corresponds to the word *attention*. Now instead of having your talking telephone ringer say *assistance*, this new binary code will make the Digitalker say the word attention when the ring signal is detected. Digitalker provides over 200 words to choose from. *Attention* and *assistance* are just examples. To program other words, be sure to insert a glass diode at every address location needing a logic 1. Then ground the rest.

Having a telephone actually say *attention, assistance* or any of over 200 words rather than having a bell clang is indeed a far-out concept. By using the technology at hand, you can design circuits that can perform breathtaking feats of electronic magic. The best thing of all is that all this wizardry can be created by you, the electronic hobbyist.

HOW THE DIGITALKER WORKS

You have been given the fast shuffle in the explanation of the operation of the Digitalker. But this omission is to be rectified immediately.

The biggest problem that designers of speech synthesizers had to cope with is the synthetic or artificial sound of the derived output. Basically because most speech chips made use of *phoneme* (the elements that all words are made of). By stringing a number of phonemes together, an infinite number of words can be created. But the speech created is hard to understand unless you listen very carefully.

The National Semiconductor Digitalker produces speech in a totally different way—by using what is called *delta modulation*. Using this modulation process, a higher quality of speech can be produced while using the least amount of memory storage.

But there is also a problem with this scheme. Only words selected by National Semiconductor can be spoken by the Digitalker. This is true because the complex speech patterns are already stored in the memory of four ROM chips (SSR1, SSR2, SSR5, and SSR6). Also, the ROM chips must be used in pairs (or they can be used separately). For example, the first 72 address locations (00000000 to 01000111) are contained on ROM SSR1 and on ROM SSR5. The remaining 72 addresses (01001000 to 10001111) are located on ROM SSR2 and ROM SSR6.

So if there are words you wish the Digitalker to say, and they are all located within the first 72 addresses of SSR1 and SSR5, all that would be produced is unintelligible, invalid speech because these two ICs already have their own words programmed at the same address. You will be programming the Digitalker to say one word, but in reality there are two words that the circuit will be trying to say.

To get the desired words to be spoken, some ROM manipulation can take place. Table 15-1 shows the legal ROM interchange that can take place. But remember, if you deviate from this listing, your speech circuit will fail to create understandable words.

Table 15-1. The Digitalker can be purchased with four ROMs. To some degree, they can be interchanged with each other.

ROM Interchange Table	
Digitalker(tm) ROM	Interchange With
SSR1	Stand Alone
SSR2	Stand Alone
SSR5	Stand Alone
SSR6	Stand Alone
SSR1	SSR2
SSR5	SSR6
SSR1	SSR6
SSR2	SSR5

Any other ROM combinations will result in an unintelligible speech output.

So far, you read that a binary code corresponds to words that are to be spoken by the Digitalker. Table 15-2 is the master list for the words programmed into the SSR1 and SSR2 memory chips, and Table 15-3 lists the binary equivalents for an additional 130 words programmed into the SSR5 and SSR6 memory chips.

Table 15-2. The master word list for the SSR1 and the SSR2 Digitalker ROMs.

Digitalker Master Word List (SSR1/SSR2)

Word	Address SW8	SW1	Word	Address SW8	SW1	Word	Address SW8	SW1
This is Digitalker	0000	0000	Q	0011	0000	IS	0110	0000
ONE	0000	0001	R	0011	0001	IT	0110	0001
TWO	0000	0010	S	0011	0010	KILO	0110	0010
THREE	0000	0011	T	0011	0011	LEFT	0110	0011
FOUR	0000	0100	U	0011	0100	LESS	0110	0100
FIVE	0000	0101	V	0011	0101	LESSER	0110	0101
SIX	0000	0110	W	0011	0110	LIMIT	0110	0110
SEVEN	0000	0111	X	0011	0111	LOW	0110	0111
EIGHT	0000	1000	Y	0011	1000	LOWER	0110	1000
NINE	0000	1001	Z	0011	1001	MARK	0110	1001
TEN	0000	1010	AGAIN	0011	1010	METER	0110	1010
ELEVEN	0000	1011	AMPERE	0011	1011	MILE	0110	1011
TWELVE	0000	1100	AND	0011	1100	MILLI	0110	1100
THIRTEEN	0000	1101	AT	0011	1101	MINUS	0110	1101
FOURTEEN	0000	1110	CANCEL	0011	1110	MINUTE	0110	1110
FIFTEEN	0000	1111	CASE	0011	1111	NEAR	0110	1111
SIXTEEN	0001	0000	CENT	0100	0000	NUMBER	0111	0000
SEVENTEEN	0001	0001	400HZ TONE	0100	0001	OF	0111	0001
EIGHTEEN	0001	0010	80HZ TONE	0100	0010	OFF	0111	0010
NINETEEN	0001	0011	20MS SILENCE	0100	0011	ON	0111	0011
TWENTY	0001	0100	40MS SILENCE	0100	0100	OUT	0111	0100
THIRTY	0001	0101	80MS SILENCE	0100	0101	OVER	0111	0101
FORTY	0001	0110	160MS SILENCE	0100	0110	PARENTHESIS	0111	0110
FIFTY	0001	0111	320MS SILENCE	0100	0111	PRECENT	0111	0111
SIXTY	0001	1000	CENTI	0100	1000	PLEASE	0111	1000
SEVENTY	0001	1001	CHECK	0100	1001	PLUS	0111	1001
EIGHTY	0001	1010	COMMA	0100	1010	POINT	0111	1010
NINETY	0001	1011	CONTROL	0100	1011	POUND	0111	1011
HUNDRED	0001	1100	DANGER	0100	1100	PULSES	0111	1100
THOUSAND	0001	1101	DEGREE	0100	1101	RATE	0111	1101
MILLION	0001	1110	DOLLAR	0100	1110	RE	0111	1110
ZERO	0001	1111	DOWN	0100	1111	READY	0111	1111
A	0010	0000	EQUAL	0101	0000	RIGHT	1000	0000
B	0010	0001	ERROR	0101	0001	SS (Note 1)	1000	0001
C	0010	0010	FEET	0101	0010	SECOND	1000	0010
D	0010	0011	FLOW	0101	0011	SET	1000	0011
E	0010	0100	FUEL	0101	0100	SPACE	1000	0100
F	0010	0101	GALLON	0101	0101	SPEED	1000	0101
G	0010	0110	GO	0101	0110	STAR	1000	0110
H	0010	0111	GRAM	0101	0111	START	1000	0111
I	0010	1000	GREAT	0101	1000	STOP	1000	1000
J	0010	1001	GREATER	0101	1001	THAN	1000	1001
K	0010	1010	HAVE	0101	1010	THE	1000	1010
L	0010	1011	HIGH	0101	1011	TIME	1000	1011
M	0010	1100	HIGHER	0101	1100	TRY	1000	1100
N	0010	1101	HOUR	0101	1101	UP	1000	1101
O	0010	1110	IN	0101	1110	VOLT	1000	1110
P	00101	111	INCHES	0101	1111	WEIGHT	1000	1111

NOTE 1: "SS" makes any single word plural
NOTE 2: Address 143 (WEIGHT) is the last legal address in this particular word list.
   Exceeding address 143 will produce pieces of unintelligible invalid speech data.

Table 15-3. The master word list for the SSR5 and the SSR6 Digitalker ROMs.

Digitalker Master Word List (SSR5/SSR6)

Word	Address SW8	SW1	Word	Address SW8	SW1	Word	Address SW8	SW1
ABORT	0000	0000	FIRE	0011	0000	RECEIVE	0110	0000
ADD	0000	0001	FIRST	0011	0001	RECORD	0110	0001
ADJUST	0000	0010	FLOOR	0011	0010	REPLACE	0110	0010
ALARM	0000	0011	FORWARD	0011	0011	REVERSE	0110	0011
ALERT	0000	0100	FROM	0011	0100	ROOM	0110	0100
ALL	0000	0101	GAS	0011	0101	SAFE	0110	0101
ASK	0000	0110	GET	0011	0110	SECURE	0110	0110
ASSISTANCE	0000	0111	GOING	0011	0111	SELECT	0110	0111
ATTENTION	0000	1000	HALF	0011	1000	SEND	0110	1000
BRAKE	0000	1001	HELLO	0011	1001	SERVICE	0110	1001
BUTTON	0000	1010	HELP	0011	1010	SIDE	0110	1010
BUY	0000	1011	HERTZ	0011	1011	SLOW	0110	1011
CALL	0000	1100	HOLD	0011	1100	SLOWER	0110	1100
CAUTION	0000	1101	INCORRECT	0011	1101	SMOKE	0110	1101
CHANGE	0000	1110	INCREASE	0011	1110	SOUTH	0110	1110
CIRCUIT	0000	1111	INTRUDER	0011	1111	STATION	0110	1111
CLAER	0001	0000	JUST	0100	0000	SWITCH	0111	0000
CLOSE	0001	0001	KEY	0100	0001	SYSTEM	0111	0001
COMPLETE	0001	0010	LEVEL	0100	0010	TEST	0111	0010
CONNECT	0001	0011	LOAD	0100	0011	TH (NOTE 2)	0111	0011
CONTINUE	0001	0100	LOCK	0100	0100	THANK	0111	0100
COPY	0001	0101	MEG	0100	0101	THIRD	0111	0101
CORRECT	0001	0110	MEGA	0100	0110	THIS	0111	0110
DATE	0001	0111	MICRO	0100	0111	TOTAL	0111	0111
DAY	0001	1000	MORE	0100	1000	TURN	0111	1000
DECREASE	0001	1001	MOVE	0100	1001	USE	0111	1001
DEPOSIT	0001	1010	NANO	0100	1010	UTH (NOTE 3)	0111	1010
DIAL	0001	1011	NEED	0100	1011	WAITING	0111	1011
DIVIDE	0001	1100	NEXT	0100	1100	WARNING	0111	1100
DOOR	0001	1101	NO	0100	1101	WATER	0111	1101
EAST	0001	1110	NORMAL	0100	1110	WEST	0111	1110
ED (NOTE 1)	0001	1111	NORTH	0100	1111	SWITCH	0111	1111
ED (NOTE 1)	0010	0000	NOT	0101	0000	WINDOW	1000	0000
ED (NOTE 1)	0010	0001	NOTICE	0101	0001	YES	1000	0001
ED (NOTE 1)	0010	0010	OHMS	0101	0010	ZONE	1000	0010
EMERGENCY	0010	0011	ONWARD	0101	0011	NOTE 4		
END	0010	0100	OPEN	0101	0100			
ENTER	0010	0101	OPERATOR	0101	0101			
ENTRY	0010	0110	OR	0101	0110			
ER	0010	0111	PASS	0101	0111			
EVACUATE	0010	1000	PER	0101	1000			
EXIT	0010	1001	PICO	0101	1001			
FAIL	0010	1010	PLACE	0101	1010			
FAILURE	0010	1011	PRESS	0101	1011			
FARAD	0010	1100	PRESSURE	0101	1100			
FAST	0010	1101	QUARTER	0101	1101			
FASTER	0010	1110	RANGE	0101	1110			
FIFTH	0010	1111	REACH	0101	1111			

NOTE 1: "ED" is a suffix that can be used to make any present tense word become a past tense word. The way we say "ED", however does vary from one word to the next. For that reason, N/S has offered 4 different sounds. It is suggested that each be tested with the desired word for best results. Address 31 or 32 should be used with words ending in "T" or "D", such as exit or load. Address 34 should be used with words ending with soft sounds such as ask. Address 33 should be used with all other words.

NOTE 2: "TH" is a suffix that can be added to words like six, seven, eight to form adjetive words like sixth, seventh, eight.

NOTE 3: "UTH" is a suffix that can be added to words like twenty & forty to form adjective like thirtiehth, fortieth, ect.

NOTE 4: Address 130 is the last legal address in this word list. Exceeding this address will produce pieces of unintelligible invalid speech data.

Just by looking up a word in Tables 15-2 and 15-3 and placing the binary code associated with that word at the data input of the speech synthesizer, you can easily command the Digitalker to vocalize.

For all of you who are interested in the internal operation of the Digitalker, refer to Fig. 15-3 (a block diagram). Figure 15-4 (the pinout of the speech synthesizer), and Fig. 15-5 (the pinout of the associated ROM memory chips).

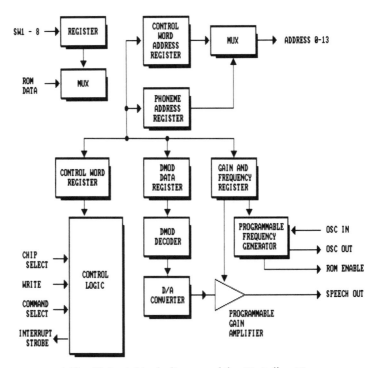

Fig. 15-3 *A block diagram of the Digitalker IC.*

HANDLING THE DIGITALKER

The Digitalker makes use of a technology called N-channel MOS (metal-oxide semiconductor) integration. Chips using this technology are prone to internal failures and breakdowns from static discharge and of course, incorrect circuit wiring.

An expert cannot look over your shoulder when you are wiring the circuit, there are a number of procedures that can minimize the risk of static electricity. First, before working on the circuit, you must have a way of discharging any electrical charge you might have in your body. A static charge can be picked up just by taking a stroll across a carpeted room. We have to get rid of this destructive element before proceeding.

Figure 15-6 shows a common solution to this all too familiar hazard. It is a *grounding strap*. The electrical conductive strap is placed around your wrist, and the wire dangling from the band is then placed in the ground side of an

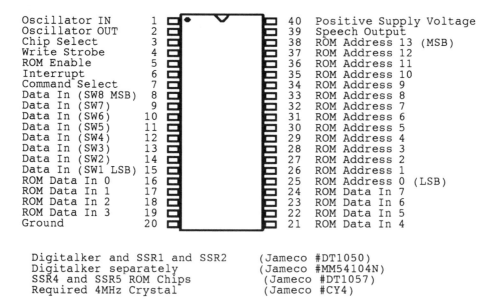

Oscillator IN	1		40	Positive Supply Voltage
Oscillator OUT	2		39	Speech Output
Chip Select	3		38	ROM Address 13 (MSB)
Write Strobe	4		37	ROM Address 12
ROM Enable	5		36	ROM Address 11
Interrupt	6		35	ROM Address 10
Command Select	7		34	ROM Address 9
Data In (SW8 MSB)	8		33	ROM Address 8
Data In (SW7)	9		32	ROM Address 7
Data In (SW6)	10		31	ROM Address 6
Data In (SW5)	11		30	ROM Address 5
Data In (SW4)	12		29	ROM Address 4
Data In (SW3)	13		28	ROM Address 3
Data In (SW2)	14		27	ROM Address 2
Data In (SW1 LSB)	15		26	ROM Address 1
ROM Data In 0	16		25	ROM Address 0 (LSB)
ROM Data In 1	17		24	ROM Data In 7
ROM Data In 2	18		23	ROM Data In 6
ROM Data In 3	19		22	ROM Data In 5
Ground	20		21	ROM Data In 4

```
Digitalker and SSR1 and SSR2      (Jameco #DT1050)
Digitalker separately             (Jameco #MM54104N)
SSR4 and SSR5 ROM Chips           (Jameco #DT1057)
Required 4MHz Crystal             (Jameco #CY4)
```

Fig. 15-4 *Pinout assignments for the Digitalker.*

ROM Address 7	1		24	Positive Supply Voltage
ROM Address 6	2		23	ROM Address 8
ROM Address 5	3		22	ROM Address 9
ROM Address 4	4		21	ROM Address 12
ROM Address 3	5		20	Chip Select
ROM Address 2	6		19	ROM Address 10
ROM Address 1	7		18	ROM Address 11
ROM Address 0	8		17	ROM Data 8
ROM Data 1	9		16	ROM Data 7
ROM Data 2	10		15	ROM Data 6
ROM Data 3	11		14	ROM Data 5
Ground	12		13	ROM Data 4

```
SSR1 and SSR2 are included with the Digitalker IC (Jameco #DT1050)
  SSR4 and SSR5 can be purchased separately (Jameco #DT1057)
```

Fig. 15-5 *Pinout for the Digitalker ROM chips.*

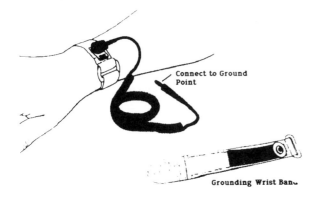

Connect to Ground
Point

Grounding Wrist Band

Fig. 15-6 *Before handling the Digitalker IC, be sure you have grounded yourself. This prevents a static buildup that might destroy the delicate IC.*

electrical outlet. The wire can be connected to a cold-water pipe. By using this device, all electrical charges generated by your movements can be dissipated harmlessly to ground through the resistance of the wrist band.

Another method of preventing the static demons from ruining a precious piece of electronics is to use a soldering iron or pencil with a grounded tip. To see if your soldering iron has a grounded tip, look at the ac plug of the iron. If it has a third prong, your iron is grounded. If not, a grounded tip might be a wise investment to make before continuing any further on the project.

Another avenue to explore is the use of IC sockets for both the Digitalker and its associated ROM chip. The Digitalker is a 40-pin monster. If you are not familiar with handling such a beast, the use of IC sockets are strongly recommended. They not only protect the ICs from the heat of soldering, but if you inadvertently burn out the chips, just pull the IC from the socket and replace it with another. On the other hand, if no sockets were used, you will have one heck of a time unsoldering 24 or even 40 wires from the IC to replace the defective component. Although there are a number of different designs on the market (PC mount and wire wrap), they all have the goal of protecting your IC investment. Sockets can be purchased at a very reasonable cost at just about any local or mail order house. It is well worth the investment.

CONSTRUCTING THE TALKING TELEPHONE RINGER

The construction of the first talking telephone ringer is not critical because you are working only with audio frequencies. So the use of a perf board will be just fine. As you can tell by now, PCBs (printed circuit boards) are highly recommended for all these designs. They not only dress up the finished project, the PCB also provides the circuit a higher degree of reliability. With one look at the schematics, you might seem intimidated by its complexity and shun the idea of using a PC board. Especially if you have to design one on your own, you have a reason to be intimidated. This is quite a complicated circuit. So if you are not familiar with the procedure necessary to design double-sided PC board artwork, do not make an attempt. You will be in for an awful experience.

So now that you are probably discouraged from using a printed circuit board, turn your attention to your auxiliary building medium: the perf board.

The Digitalker and its associated ROM memory are relatively expensive items. Any mis-wires in the circuit can spell disaster for the chips. So the use of IC sockets for both the Digitalker and its ROM are strongly recommended.

As mentioned, IC sockets come in two varieties: PC mount and wire wrap. If you plan to make use of wire wrapping to assemble the circuit, sockets making use of extra-long connection pins should be used. These speciality sockets allow a number of wires to be connected to each pin with room to spare. PC mount sockets should not even be considered for use with this complicated circuit, because the pins are too short. This problem will prevent the soldering of more than two wires per pin. So invest a few pennies more and buy the wire-wrap IC sockets.

Another drawback to a circuit of this type is using ICs that have so many pins. For instance, Digitalker has 40 connecting points, and the memory chips have 24 points. To count each and every pin one by one every time you want to solder in a wire in the circuit will compound the complexity tenfold. It also will provide additional avenues for wiring errors. You really do not need errors in a circuit such as this.

To eradicate this ever-present predicament, one versatile manufacturer has come up with a simple and obvious solution. The solution is to cut rectangular pieces of plastic and print out the pin numbers of the IC in a clockwise direction (see Fig. 15-7). These plastic strips are gently impaled on the pins of a wire-wrap IC socket.

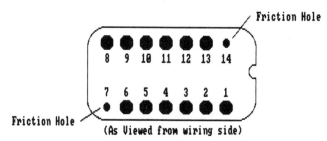

Fig. 15-7 *To help prevent mis-wires, manufacturers have designed IC markers. These markers have the IC pin number printed right on the plastic.*

It is said that a picture is worth a thousand words, so turn your attention to Fig. 15-8. This illustration depicts how wire-wrap ID markers are installed and used with a 14-pin socket mounted on a typical perf board. This figure illustrates a 14 pin IC marker, but they also come in 8-, 16-, 24- and even 40-pin models. So by making use of these inexpensive, innovative pieces of plastic, the risk of miscounting IC pin numbers are brought down to almost zero.

To continue the discussion on the assembly of the talking telephone ringer, components such as the diodes, ICs, and one electrolytic capacitor (C10) are polarity-sensitive devices and can only be installed one way. Deviating from this scheme can result in an inoperable project and many hours of unnecessary troubleshooting time.

Resistor R6 (50kΩ) is the main volume control. It is used to control the amount of audio being fed into the power amplifier (IC3). For this reason, R6 should be a front-panel control. To get a smooth and uniform increase (as well

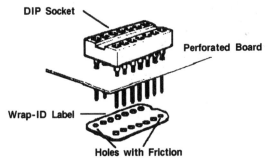

Fig. 15-8 *Using the IC markers (wrap ID label).*

as decrease) in volume, R6 should be an audio-taper potentiometer. Just as other components come in a wide variety of makes and models, potentiometers are no exception. Physical sizes, resistance values, wattages as well as price should also be considered when purchasing volume controls.

TESTING AND FINAL ASSEMBLY

When your talking telephone ringer has been completed, check—then check again—all wiring to make sure that there are no mis-wires as well as solder bridges, solder splashes or bits of copper wires imbedded somewhere on the board. This might seem a bit trivial, but a small piece of wire in the wrong place can pop a $30.00 IC in a wink of an eye. So be warned.

When your final inspection of the board is complete and you are satisfied all mis-wires have been corrected, connect a 7-to-11 Vdc wall transformer to the circuit as shown in Fig. 15-1. Using a voltmeter, test for the proper 5 V output at IC1 (LM7805). And make sure the same 5 Vdc supply is being delivered to pin 24 of the ROM memory (IC4). When tested and verified, now (again using your voltmeter) check for the proper 7 to 11 Vdc voltage levels at pin 4 of IC2, pin 6 of IC3, and pin 40 of IC5. Only when all voltages are within normal levels, you should consider installing the ICs.

With the wall transformer disconnected from the ac outlet and yourself grounded with a grounding strap, remove the Digitalker from its protective plastic case. Carefully install the chip into the 40-pin IC socket, be sure to maintain proper orientation and be sure that no pins are bent under the body as the component is inserted. Next, remove the ROM chip from its shipping container and install it on the 24-pin IC socket. Again be careful for proper orientation and bent pins.

But which ROM do I install? I have four. Figure 15-2 shows the wiring needed to command the Digitalker to say the word *attention* or *assistance*. By looking up these two words on the master word list (Table 15-2 and Table 15-3), you will soon see that these two words are located on ROM memory SSR5 and SSR6. To be precise, the two words are located on SSR5. It is this IC that is installed in the IC4 socket.

When all other integrated circuits have been installed on the board, it is time to test this baby out. Using a standard telephone line cord, connect the green wire from the cord to the free end of the resistor R2. Connect the red wire to resistor R1. When complete, plug the modular plug into a wall jack. Then apply power.

If all is operating, your talking telephone project should come to life as soon as you put the transformer into the wall socket. It should say either *assistance* or *attention*. If you hear this, so far, so good.

Now is the time to have that friend call you up on the phone. When the ringing signal is present across the telephone line, the output of IC1 will go low, and the output of IC2 will go high. It is this high that will instruct the talking telephone ringer to speak.

Let the phone ring a few times, Digitalker will repeat the same word over and over again until you answer the telephone or the calling party hangs up.

Digitalker will say the same word over and over again? Seems kind of dull and boring after a while. Well, seeing that you have the main Digitalker wired and operational, maybe you can make a few changes to the interface and make it a little bit more sophisticated.

TALKING TELEPHONE RINGER (CIRCUIT 2)

Figure 15-9 shows one circuit modification. From this illustration, you can see that a 4013 (dual D flip/flop) and the now-illustrious 74188 PROM chip have been added.

The operation of this added circuitry is quite simple. The 4013 is wired as a two-bit counter. The dual D flip-flop counts off the ring signals as they are detected. The output data, in binary form, is presented at pins 1 and 3.

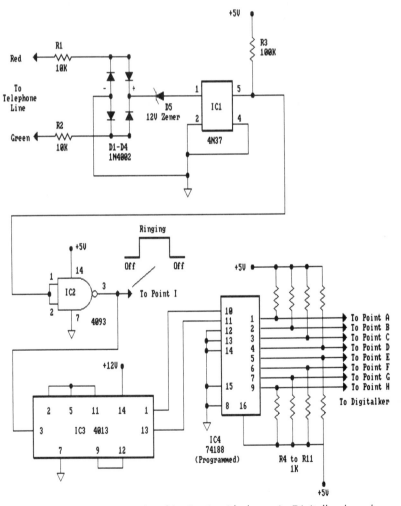

Fig. 15-9 *By incorporating this circuit with the main Digitalker board, you can have your project speak a number of words every time a ring signal is detected.*

Table 15-4 shows exactly what happens with every detected ring signal. At power up, let's assume that the output binary code is 0 0. When the first ring is detected, the 4013 changes state to produce an output of 0 1. When the second signal is received, the 4013 changes again—this time to produce an output of 1 0. The third signal creates an output of 1 1. And finally, the fourth signal detected will reset the flip-flop counter to its original, or resting, position of 0 0.

If you use these binary outputs as addresses for a 74188 PROM chip, you have the makings of a circuit that will change its output to a predetermined sequence at a given input. In other words, if you program the eight outputs of the 74188 to get a valid Digitalker address for each detected ring signal, you can make your basic talking telephone ringer say any one of four selected words. (See Table 15-5.)

Table 15-4. The 4013 truth table.

4013 Truth Table

Ring Count	Output	
	Pin 1	Pin 13
Power Up	0	0
First Pulse	0	1
Second Pulse	1	0
Third Pulse	1	1
Fourth Pulse		RETURN

Table 15-5. Talking telephone ringers truth table.
This table is to be programmed into a 74188 PROM IC.

Truth Table for the Talking Telephone Ringer

Thumbwheel Display	Binary Code Input SW7 SW6 SW5 SW4 SW3					Memory Chip Programmer SW8 Setting B7 B6 B5 B4 B3 B2 B1 B0							
00	0	0	0	0	0	0	0	1	1	1	0	0	1
01	0	0	0	0	1	0	0	0	0	1	0	0	0
02	0	0	0	1	0	0	0	1	1	1	0	0	1
03	0	0	0	1	1	0	0	0	0	0	1	1	1

With the following Truth Table programmed into a 74188 chip, Digitalker(tm) will say the following words every time the telephone rings:

```
Code 0 0 0 0 0 ....... "HELLO"
Code 0 0 0 0 1 ....... "ATTENTION"
Code 0 0 0 1 0 ....... "HELLO"
Code 0 0 0 1 1 ....... "ASSISTANCE"
```

NOTE: SSR5 ROM Memory Chip must be installed with Digitalker to say the words indicated above.

With the PROM programmer switches set to binary 0 0 0 0, you can "blow out" the internal fuses of the 74188 at positions B0, B3, B4, and B5. If you look up this code on the Master Word List in Table 15-3, you will see that this output corresponds to the word *HELLO*.

If you set PROM programmer to its "03," or 0 0 0 1 1, position, you can blow out the fuses for B0, B1, and B2. This code now corresponds to the word *ASSISTANCE*. If you have taken the time to program all the possible binary addresses from the 4013 IC, your circuit will say:

- *HELLO*, on the first ring
- *ATTENTION*, on the second ring
- *HELLO*, on the third ring
- *ASSISTANCE*, on the fourth ring

These are not the only words you can program into the PROM. Just look at the Master Word List and select the desired output. Then program their corresponding addresses into the 74188 IC. Of course, you must have the associated Digitalker ROM memory placed into the IC4 socket. (Truth tables are in Tables 15-4 and 15-5.)

Another consideration is if you want the ringer to say a word located in SSR1 and a second word located in SSR5, all you have to do is to wire another 24-pin IC socket in parallel with the existing IC4 and call in IC6. For example, solder a wire to pin 1 of IC4. The other end of the same wire should be soldered to pin 1 of IC6 (new IC socket). Just continue soldering wires in parallel until all 24 pins are soldered (remember to install a new 0.1 μF capacitor from pin 24 to ground of the new IC socket).

PROGRAMMING THE PROM

To help provide a foolproof method of programming the PROM, step-by-step instructions are included. So read each line very carefully. Double check all switch positions before hitting that program button. And of course, verify all logic 1 levels.

1. Flip the thumbwheel to display a 00 or flip SW3, SW4, SW5, SW6, and SW7 to their logic 0 positions. Set SW8 to the B0 position; then press SW1 to program. Verify a logic 1 at this address.
2. Still at address 00, set SW8 to the B3 position. Press SW1 to program.
3. Still at address 00, set SW8 to B4. Press SW1 to program.
4. Still at address 00, set SW8 to B5. Press SW1 to program.
5. Flip the thumbwheel to display a 01 or flip SW3 to its up position. Set SW8 to B3. Press SW1 to program.
6. Flip the thumbwheel to display a 02 or flip SW4 to its up position. Set SW8 to B1. Press SW1 to program.
7. Still at address 02, set SW8 to B3. Press SW1 to program.

8. Still at address 02, set SW8 to B4. Press SW1 to program.
9. Still at address 02, set SW8 to B5. Press SW1 to program.
10. Flip the thumbwheel to display a 03 or flip SW3 and SW4 to their up positions. Set SW8 to B0. Press SW1 to program.
11. Still at address 03, set SW8 to B1. Press SW1 to program.
12. Still at address 03, set SW8 to B2. Press SW1 to program.

BUILDING THE NEW TALKING TELEPHONE RINGER

As stated earlier, you already have the bulk of the talking telephone ringer already assembled and troubleshot. All that remains is to wire the new interface to the main Digitalker circuit. Unlike previously mentioned, programming diodes (Fig. 15-2) have now been replaced with the 74188 PROM chip. Basically, the wiring is the same. Just solder points A to H to their corresponding points of the Digitalker main board. The output from IC2 (pin 3) remains connected to point I.

All other building precautions as mentioned previously should also be used with this interface. Just remember that when soldering this new interface into place, the dc wall transformer must be removed from the ac line. Also, it is a good idea to remove the Digitalker and its ROM memory from its sockets (use the grounding strap when removing the ICs).

TESTING THE NEW TALKING TELEPHONE RINGER

With your new talking telephone ringer connected to the telephone line through a standard modular cord, plug in the wall transformer and retest for proper voltage levels. When all seems to be acceptable, remove the transformer from the wall and reinsert both the Digitalker and the ROM chip into its proper IC socket (make sure that the delicate IC pins have not been bent under the body of the component and be sure you have the proper orientation). Then apply power.

Now with that friend calling your number again, this time, instead of your talking ringer saying only one word, you will hear a series of four different words. Each being spoken with the detection of every ring signal.

PURCHASING THE DIGITALKER

You are probably thinking that the Digitalker is a specialty device and not available to just any hobbyist. That statement cannot be further from the truth. To the contrary, the Digitalker and its ROM chips can be purchased from a mail order electronics organization that you have probably dealt with before:

Jameco Electronics
1355 Shoreway Road
Belmont, CA 94002
(415) 592-8097

The Digitalker and its ROM memory chips (SSR1 and SSR2) can be purchased under the Jameco part number DT1050. The Digitalker itself is being sold under the number MM54104N. (Note: This part number does not include the needed SSR1 and SSR2 ROM chips.) The memory expansion chips (otherwise known as the SSR5 and SSR6) have the Jameco part number DT1057.

When buying the individual chips or in sets of three, you might also want to consider purchasing the special 4 MHz crystal used as a system timer in the Digitalker circuit. This crystal can also be purchased from Jameco under the part number of CY4.

Availability and pricing of this speech and memory chips are unavailable at the time of this writing. But for a complete listing of parts and pricing, just drop Jameco a note and request their most recent catalog.

PARTS LIST FOR THE MAIN DIGITALKER CIRCUIT (FIG. 15-1)

R1	1.5kΩ resistor
R2	1MΩ resistor
R3	620kΩ resistor
R4	9.1kΩ resistor
R5	10kΩ resistor
R6	50kΩ audio taper potentiometer
R7	10 Ω resistor
C1 C4 C5	
C6 C7 C8	
C11	0.1 μF disk capacitor
C2 C3	50 pF disk capacitor
C9	0.05 μF disk capacitor
C10	20 μF, 10 V electrolytic capacitor
IC1	LM7805 5 V voltage regulator
IC2	LM346 operational amplifier
IC3	LM386 power amplifier
IC4	Speech ROM chip(s)*
IC5	Digitalker IC*
XTAL1	4 MHz crystal*
1	dc wall transformer, 6 to 11 Vdc 500 mA
1	Perforated board
1	8 Ω speaker

*These parts can be purchased from Jameco Electronics, 1355 Shoreway Road, Belmont, CA 94002.

PARTS LIST FOR THE TALKING TELEPHONE RINGER INTERFACE (FIG. 15-2)

R1 R2	10 kΩ resistors
R3	100 kΩ resistor
D1 to D4	1N4002 or equivalent
Programming Diodes	1N914 or equivalent
IC1	4N37 opto-isolator
IC2	4093 NAND gate Schmitt trigger
1	Perforated board
1	Modulator telephone line cord

PARTS LIST FOR THE TALKING TELEPHONE RINGER (FIG. 15-9)

R1 R2	10 kΩ resistors
R3	100 kΩ resistor
R4 to R11	1 kΩ resistors
D1 to D4	1N4002 or equivalent
D5	12 V Zener diode
IC1	4N37 Opto-isolator IC
IC2	4093 NAND gate Schmitt trigger
IC3	4013 Dual D-type flip/flop
1	Perforated board
1	Standard telephone line cord

☎16
The Talking Telephone

Chosen to one of the top ten designs of 1989 by:

MANY TELEPHONE PROJECTS HAVE GRACED THE PAGES OF MANY FINE BOOKS AND magazines throughout the years. Once assembled, tested, and installed, the novelty of the devices often wears off within weeks or even days. The circuit that required many hours of assembly and a certain amount of cash is thrown into the familiar junk box. There it might lie around waiting to be cannibalized for parts that can be used on future projects.

Please allow this book to put an end to the impulse of assembling telephone projects that have a life expectancy of about 10 hours. Presented in this chapter is a circuit that is called the talking telephone. Once assembled and properly connected to the telephone line, the talking telephone will actually repeat (through the use of National Semiconductor's Digitalker speech synthesizer) the number as you are dialing. The talking telephone will not only be the topic of conversation at home or at the office, but it will prevent mis-dialed telephone numbers by allowing you to hear the number as it is being entered but before the central office equipment bills you for the call.

A telephone enhancement device of this type is not only indispensable in the office or home but it also can be used as a tool to help visually handicapped persons perform the everyday task of successfully dialing a tone telephone without the worry of being charged for wrong numbers.

THE THEORY OF THE TALKING TELEPHONE

In the home or office, there are two main types of telephone instruments: the rotary or pulse dial and the tone dial. Because tone-dial telephones are more widely used than the rotary dial, the talking telephone is designed around tone-dial instruments. Telephones such as the 2500, 3554, and 2554, which can be found in almost any home, produce a special kind of dialing signal called dual-tone multifrequency, or DTMF for short.

By using an unusual integrated circuit called a DTMF receiver, you can convert these tones into usable binary codes that can then be fed into the data input of the Digitalker IC. It is these binary codes that are literally read by the Digitalker and converted into words like *one, two, three,* and so on. This specialty IC is called the G8870 DTMF receiver. It is the IC used on so many other projects in this book.

For a moment, go back and review the dialing characteristics of a tone telephone. By reviewing exactly what happens inside a 2500 telephone, you will better understand the complexities of the talking telephone.

A tone-dial telephone, such as the 2500, produce a dual-tone or two-tone output whose frequencies are determined by the row and column of the button pressed. As mentioned in earlier chapters, if the button 8 were pressed, the dial would produce an output consisting of 852 Hz and 1336 Hz tones. Pressing button 4 would produce an output consisting of the sum of 770 Hz and 1209 Hz tones.

When first introduced back in the 1960s, tone dials were made from comparatively large *inductors* (coils) and capacitors. These devices produced pure sine-wave outputs. Today, DTMF keypads use crystal-controlled ICs that generate the synthesized stair step wave forms. These wave forms are crude approximations of pure sine wave signals, but devices such as the G8870 can receive and decode each and every tone into its corresponding binary output. It is this output that is converted by an additional integrated circuit (IC2/74188 programmed PROM) into the needed digital code that is delivered to the input of the Digitalker. Each 8-bit input being transported to IC3 (which is programmed into IC2) selects the word, or in the case of the talking telephone, the number Digitalker is to say when any button on the tone dial has been pressed.

HOW IT TALKS

The talking telephone uses an easily obtainable speech synthesizer made by National Semiconductor. This IC, named Digitalker, stores complete words in its support ROM chips (SSR1, SSR2, SSR5, and SSR6). As mentioned in chapter 15, the three-chip Digitalker basic kit can be purchased from Jameco Electronics.The MM54104 SPC is the heart of the Digitalker, but the ROM chips form the brain. Digitalker is a 40-pin DIP integrated circuit. It has eight data lines where the binary code that corresponds to the word you wish spoken is connected. For the talking telephone, you need to have Digitalker say the numbers *one, two,* . . . *eight, nine* and the word *zero.* This is by no means the only vocabulary programmed into the available four memory chips. If you wish to review the complete listing of all the words the Digitalker is able to say, just flip to Table 15-2 and Table 15-3. By examining these two tables, you will notice that the wording required by your talking telephone circuit is contained within the SSR1 ROM. (This IC is part of the standard Digitalker IC kit available from Jameco Electronics.) The vocabulary contained within the SSR2 ROM is not needed with this project, so put the IC in a safe place; it will not be installed as part of the project. You might want to use it later for another talking circuit.

PUTTING IT ALL TOGETHER

So far, you have an IC that converts the DTMF signal of a tone-dial telephone into a binary code and an IC (Digitalker) that takes this code as an input and transforms it into the spoken word. But at a component level, how is it done? For the following, refer to the schematic diagram of the talking telephone in Figs. 16-1 and 16-2.

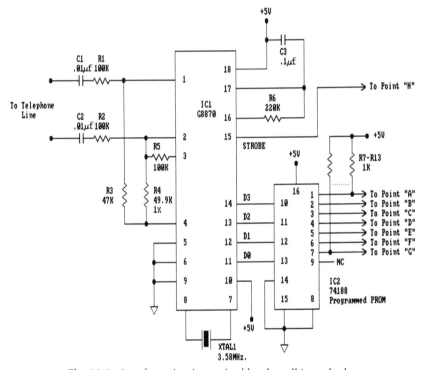

Fig. 16-1 *Interface circuit required by the talking telephone.*

Pins 1 and 2 of the G8870 (IC1) is connected directly across the telephone line through two capacitors and two resistors. This input allows the DTMF signal that is generated by pressing one of the tone-dial buttons to be passed to the complex filtering and switching electronics located deep down within the plastic body of IC1. The binary code that corresponds to the button that is pressed is made available at pins 10 to 14 (IC1).

The whole idea of this project is to have the Digitalker say a predefined word when a specific binary code is placed on its data input leads (SW8 to SW1 pins 8 to 15). For an example of this, take the binary code generated when we

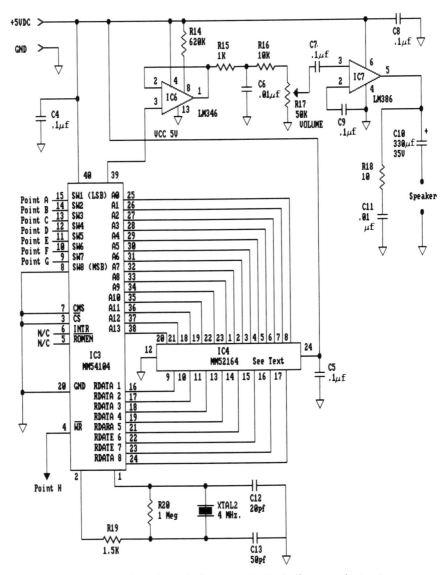

Fig. 16-2 *The schematic for the main Digitalker speech circuit.*

press the button 1 on the telephone tone dial. Table 16-1 tells you that by pressing the 1 TTP button, the generated output would be 0001. Table 16-2 (the Digitalker partial word list) informs you that if this binary code were placed on its data input, Digitalker will say the word one. So far, you have no problems. Go a little further. Assume you have pressed the button 5 on the tone dial. Table 16-1 indicates that the binary output of the G8870 will be 0101. Table 16-2 shows you that the code 0101 will make the Digitalker say the word *five*. Oh wow! This is great! Or is it? Go just a bit further. What would happen if you press the 0 button? Table 16-1 shows that the code 1010 will be generated.

Table 16-1. Frequency assignments and binary output for the G8870 DTMF IC.

Button	Low Frequency Component (Hz.)	High Frequency Component (Hz.)	HEX Output Format 3 2 1 0
1	697	1209	0 0 0 1
2	697	1336	0 0 1 0
3	697	1477	0 0 1 1
4	770	1209	0 1 0 0
5	770	1336	0 1 0 1
6	770	1477	0 1 1 0
7	852	1209	0 1 1 1
8	852	1336	1 0 0 0
9	852	1477	1 0 0 1
0	941	1336	1 0 1 0
*	941	1209	1 0 1 1
#	941	1477	1 1 0 0

Table 16-2. Partial word list for the Digitalker.

Partial Digitalker Word List

Word	8-Bit Binary Address SW8 SW1
THIS IS DIGITALKER	0 0 0 0 0 0 0 0
ONE	0 0 0 0 0 0 0 1
TWO	0 0 0 0 0 0 1 0
THREE	0 0 0 0 0 0 1 1
FOUR	0 0 0 0 0 1 0 0
FIVE	0 0 0 0 0 1 0 1
SIX	0 0 0 0 0 1 1 0
SEVEN	0 0 0 0 0 1 1 1
EIGHT	0 0 0 0 1 0 0 0
NINE	0 0 0 0 1 0 0 1
ZERO	0 0 0 1 1 1 1 1
TEN	0 0 0 0 1 0 1 0
ELEVEN	0 0 0 0 1 0 1 1
TWELVE	0 0 0 0 1 1 0 0
400 HERTZ TONE	0 1 0 0 0 0 0 1

All words listed above are contained on the SSR1 ROM Memory Chip

Well, so far so good. The Digitalker partial word list indicates that the code of 1010 will generate the word *ten*. Something is not right. At least, you do not give out telephone numbers by saying:

1-813 46 ten 576 ten

You would say:

1-813 46 zero 576 zero

or 1-813 460 5760, right? So the talking telephone also should conform to this standard. There is no reason why it should not. That is one of the jobs of IC2. This integrated circuit is a pre-programmed ROM. With this IC (74188 or 8223), you can program it to deliver the binary output of the word *zero* (00011111) every time its input code is 1010 (or the decimal 10). You can take this programming one step further. Suppose you press the * or the # button on the telephone. Table 16-1 indicates that if this were the case, the talking telephone would say the words *eleven* and *twelve* respectively. To eliminate these mis-spoken words, IC2 can be programmed to deliver the binary code for a 400 Hz tone burst (01000001) every time the * or the # button is pushed.

The second job of the 74188 (or 8223) is to correct a strange problem associated with using the G8870 IC. If you compare the pinout of the G8870 with that of the 74188 (see Fig. 16-3), you will notice that the data-output lines of the G8870 are reversed from that of the 74188. Normally, pin 10 of the 74188 is reserved for a data 0 input. At this physical location, IC1 delivers a data 3 output if you were to parallel connect all common data lines. If this were to happen, the output data of the G8870 would not be in its normal binary sequence. Thus providing an erroneous input to the Digitalker. This is an interesting problem.

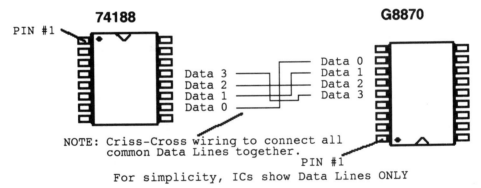

Fig. 16-3 *The physical pin relationship for the G8870 and the 74188 ICs. Note that the interconnecting wires cross each other.*

This problem can be corrected by designing a printed circuit board to compensate for it, but this would complicate the layout. Figure 16-3 shows that a crisscross arrangement would have to be instituted to allow a data 0 output from the G8870 to be connected directly to the normal D0 input of the 74188. This would also hold true for D1, D2, and the D3 bits.

So why should you go through the hassle needlessly? Just have the 74188 IC programmed with a binary code that is in reverse of the original code. That is exactly what is done. Figure 16-4 shows four parallel data lines between the G8870 and the 74188. Now the physical layout of a printed circuit board is less complicated. But remember that the output data from the G8870 is now reversed at the input of the 74188. For an example, take a closer look at Fig. 16-4.

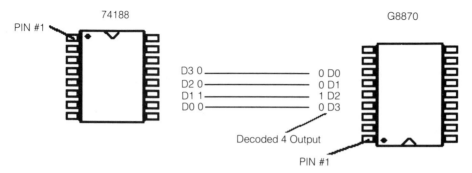

Decoded 4 Binary Output from the G8870 is 0 1 0 0. Electrical
Input to 74188 using parallel Bus as shown above is 0 0 1 0.
Binary 0 0 1 0 is equal to Decimal 2. Input to 74188 is a Mirror
Image of the G8870 output .9

Fig. 16-4 *By reversing the data-input pins of the 74188,
you can create a situation that would allow you to design a less complicated PC board.*

Pin 7 (G8870) is the normal D0 output. By connecting this data bit to pin 13 (74188), the LSB (least significant bit) from the G8870 now becomes the MSB (most significant bit) to the 74188 at the input of the PROM chip. By parallel connecting the remaining three data bits, a *mirror image* of the standard G8870 output is delivered to the 74188.

Table 16-3 is a rather complicated listing, so follow carefully and take a closer look at this rather unique programming sequence. By pressing the 1 button of the TTP of a telephone, the output from the G8870 will be 0001. This data will be reversed by the copper traces of the PC board. The results will be an input to the 74188 amounting to 1000. If you use the PROM programmer shown in appendix B, Fig. B-13, the thumbwheel setting would be the decimal 08 (the binary code for the number 08 is 1000). If you were to use the programmer shown in appendix B, Fig. B-9, the switch settings would be:

SW7 Down   SW6 Up   SW5 Down   SW4 Down   SW3 Down

    0   1   0   0   0

With the proper address settings, all that remains is to program the B0-to-B7 data-output lines of the 74188 by setting SW8 to the proper output line that will give you the needed logic 1 level.

In this example, you pressed the 1 button of the tone dial. With this button pressed, you want your circuit to say the word *one*. From the Digitalker partial word list (Table 16-2), you need an output from the PROM to read 00000001. So to say the word *one*, the programmer must blow the internal fuse of the PROM at the output location B0.

Take another example. You have just pressed the 0 (or operator) button on the TTP. The G8870 decodes the tones into the code 1010. The mirror-image input to the PROM will be 0101. The thumbwheel setting on the programmer

Table 16-3. Talking telephone truth table.
It is this table that must be programmed into a 74188 PROM IC.

Talking Telephone Truth Table			
TTP Button	G8870 Output Data D3 D2 D1 D0	74188 Input Data D3 D2 D1 D0	Programmer Thumbwheel Decimal Setting
1	0 0 0 1	1 0 0 0	08
2	0 0 1 0	0 1 0 0	04
3	0 0 1 1	1 1 0 0	12
4	0 1 0 0	0 0 1 0	02
5	0 1 0 1	1 0 1 0	10
6	0 1 1 0	0 1 1 0 (Same)	06
7	0 1 1 1	1 1 1 0	14
8	1 0 0 0	0 0 0 1	01
9	1 0 0 1	1 0 0 1 (Same)	09
OPER	1 0 1 0	0 1 0 1	05
*	1 0 1 1	1 1 0 1	13
#	1 1 0 0	0 0 1 1	03

will be 05 (address-switch setting for the Fig. B-13 programmer will be SW3 and SW5—down and SW4 and SW6—up). Then using SW8, blow the fuses at PROM output locations B0, B1, B2, B3, and B4. This will allow your circuit to say the word *zero* instead of the normal 1010 address, which would be the word *ten*.

Table 16-3 also shows that if the TTP buttons 6 and 9 were pressed, the mirror image of the binary output for these two numbers would be the same as the forward reading. So for the numbers 6 and 9, the programmer thumbwheel settings should be 06 and 09 respectively.

Go back to Figs. 16-1 and 16-2 to complete the explanation of the talking telephone. Where you left off, you have a pre-programmed IC (IC2) that takes care of two inherent problems that have been encountered and corrected. Pins 1 to 7 of IC2 deliver the appropriate binary code to the data-input line of the Digitalker (pins 8 to 15). But with this arrangement, you still have no speech output. You must provide a ground to logic 1 pulse at the write strobe of the Digitalker (pin 4). Remember the start-up pulse on the talking telephone ringer? You need a rising pulse here too. Going back to the G8870, pin 10 is a strobe output that produces the positive-going pulse every time a valid tone is detected at its input. This strobe pulse is connected directly to pin 4 of the Digitalker. Now, when any button is pressed, you have two actions taking place. The first is that the appropriate binary code is placed on the data-input line of the Digitalker. The second is that the needed ground to positive-going pulse is generated by the DTMF receiver (IC1) at pin 10.

The MM52164 seen in the schematic as IC4 is the ROM (SSR1) that contains the vocabulary needed by the talking telephone. The partial word list (Table 16-2) contains all the words (and tones) needed for the proper operation of the completed talking telephone. If you compare this list with that in Table 15-2, you will notice that all Digitalker words are located within the first half of the list. This indicates that the needed speech data is stored on the SSR1 ROM.

The SSR2 data is not needed, so this chip is omitted from the completed circuit, thus making the PC board artwork a bit less complicated.

The additional components associated with IC6 make up a filter circuit (R15, R16, and C6). IC7 is a low-voltage power amplifier IC (LM386), which has its volume controlled by potentiometer R17. This IC will provide enough volume so the spoken output of the Digitalker can be heard with little trouble.

Do not forget the power supply. Figure 16-5 illustrates a power supply that can deliver the regulated + 5 Vdc. For the proper operation of the talking telephone, two parameters must be met. The first is that the 9 Vdc wall transformer must be able to deliver at least 500 mA of current. The second is that IC5 (voltage regulator) be of the TO-3 (diamond shape) design. Also, this IC should be mounted on a heat sink. The Digitalker and its SSR1 ROM do take a lot of current to operate so you must be able to feed the current hungry circuit all it needs to operate.

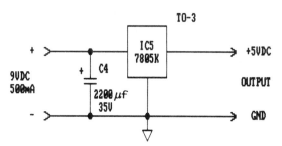

Fig. 16-5   Suggested power supply for the talking telephone.

PROGRAMMING THE PROM CHIP

For the proper operation of the talking telephone circuit, the 74188 must have the correct electronic data burned into it. The 74188 ICs cost around $1.50 (from Jameco Electronics), but repeated errors will only cost you in the long run. By following the procedure given below, blunders like programming a logic 1 where a 0 belongs will be kept to a minimum.

As in the programming of any PROM chip, double check all switch positions before hitting the program button. Also, verify all logic 1 levels. So fire-up your programmer and create an electronic brain.

1. Flip the thumbwheel to display a 08 or flip SW6 to its up position. Set SW8 to the B0 position. Press SW1 to program. Verify a logic 1 level.
2. Flip the thumbwheel to display a 04 or flip SW5 to its up position. Set SW8 to B1. Press SW1 to program. Then verify logic level.
3. Flip the thumbwheel to display a 12 or flip SW5 and SW6 to the up position. Set SW8 to B0. Press SW1 to program. Then verify logic level.
4. Still in address 12, set SW8 to B1. Press SW1 to program. Then verify logic level.

5. Flip the thumbwheel to display a 02 or flip SW4 to its up position. Set SW8 to B2. Press SW1 to program. Then verify logic level.

6. Flip the thumbwheel to display a 10 or flip SW4 and SW6 to the up position. Set SW8 to B0. Press SW1 to program. Then verify logic level.

7. Still in address 10, set SW8 to B2. Press SW1 to program. Then verify logic level.

8. Flip the thumbwheel to display a 06 or flip SW4 and SW5 to its up position. Set SW8 to B1. Press SW1 to program. Then verify logic level.

9. Still in address 06, set SW8 to B2. Press SW1 to program. Then verify logic level.

10. Flip the thumbwheel to display a 14 or flip SW4, SW5, and SW6 to the up position. Set SW8 to B0. Press SW1 to program. Then verify logic level.

11. Still in address 14, set SW8 to B1. Press SW1 to program. Then verify logic level.

12. Still in address 14, set SW8 to B2. Press SW1 to program. Then verify logic level.

13. Flip the thumbwheel to display a 01 or flip SW3 to its up position. Set SW8 to B3. Press SW1 to program. Then verify logic level.

14. Flip the thumbwheel to display a 09 or flip SW3 and SW6 to the up position. Set SW8 to B0. Press SW1 to program. Then verify logic level.

15. Still in address 09, set SW8 to B3. Press SW1 to program. Then verify logic level.

16. Flip the thumbwheel to display a 05 or flip SW3 and SW5 to the up position. Set SW8 to B0. Press SW1 to program. Then verify logic level.

17. Still in address 05, set SW8 to B1. Press SW1 to program. Then verify logic level.

18. Still in address 05, set SW8 to B2. Press SW1 to program. Then verify logic level.

19. Still in address 05, set SW8 to B3. Press SW1 to program. Then verify logic level.

20. Still in address 05, set SW8 to B4. Press SW1 to program. Then verify logic level.

21. Flip the thumbwheel to display a 13 or flip SW3, SW5, and SW6 to the up position. Set SW8 to B0. Press SW1 to program. Then verify logic level.

22. Still in address 13, set SW8 to B6. Press SW1 to program. Then verify logic level.

23. Flip the thumbwheel to display a 03 or flip SW3 and SW4 to the up position. Set SW8 to B0. Press SW1 to program. Then verify logic level.

24. Still in address 03, set SW8 to B6. Press SW1 to program. Then verify logic level.

CONSTRUCTION

Printed circuit layout patterns for both sides of the board are shown in Fig. 16-6 (component side) and Fig. 16-7 (solder side) if you wish to make your own. If you do not want to go through the trouble of making your own, Del-Phone Industries, Spring Hill, Florida, will supply the etched and drilled printed circuit board as seen in this chapter. They also sell pre-programmed 74188 PROM chips. Just drop them a line at the address given in the parts list and request current pricing.

Please note that if you want to make your own board or purchase it from Del-Phone, the board will not contain plated-through holes. You MUST solder all components on both sides of the board where needed. This is important.

The Digitalker used N-channel MOS technology. It can be damaged by static charges. To prevent a static buildup, use a 15 to 20 W grounded soldering pencil in assembly. Also, do not handle the Digitalker IC without first grounding yourself. (Remember the grounding strap in chapter 15?) Just by touching the legs, you can destroy the ICs. So observe normal precautions when handling the device. Leave the Digitalker and its SSR1 ROM chips in their protective shipping container until you are ready to install them.

Before mounting any of the ICs, install and solder all associated components such as resistors and capacitors. To help in finding the component locations on the board, see Fig. 16-8 (component parts layout diagram). Observe normal soldering precautions. Be especially careful with capacitors C4 and C10. They are electrolytics and can be installed in the circuit one way only.

Observe resistor color codes when stuffing the board. Except for resistor R4 (49.9 kΩ, 1 percent), value substitutions can be made. But keep the replacements within a reasonable value. Do not exchange a 68 kΩ resistor for a 47 kΩ resistor. You must be realistic. The only unusual component installation can be found at the output of the PROM IC (resistors R7 to R13). To conserve precious board real estate and to prevent any digital noise from disrupting the normal binary data transfer to the Digitalker, these resistors have to be mounted standing on end. Figure 16-9 illustrates this mounting procedure. Resistors R7 to R13 should first be installed and soldered in their proper location on the board. Next, bend the free lead of R13 to the left. Now solder the leads of resistors R7 to R12 to the bent lead of R13 as shown. To complete the installation, a short bus wire is to be soldered to the shorted end of this resistor network. The free end of the bus wire will be soldered to pad just adjacent to R7. This pad delivers a + 5 V to the resistors.

When all components have been installed and soldered, you can install and solder IC1, IC2, IC4, IC5, IC6, and IC7. Again, observe the orientation of the device. ICs, like electrolytic capacitors, can only be mounted one way—the correct way. When the ICs have been installed, mount the Digitalker IC (IC3).

First be sure you have grounded yourself. Pick up the IC by the edges only; avoid touching the legs. Place the IC in its proper location on the board. Note that this IC requires a number of leads to be soldered on both sides of the board.

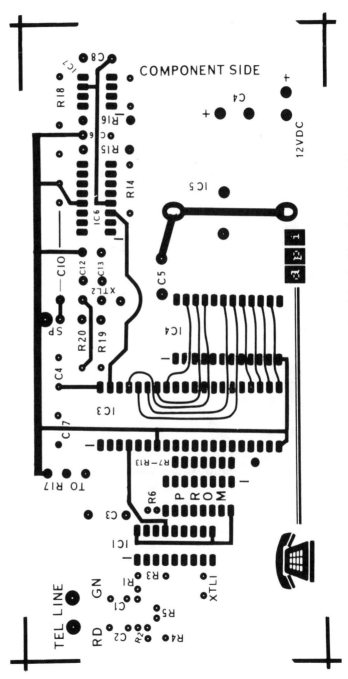

Fig. 16-6 Artwork for the component side of the double-sided PC board.

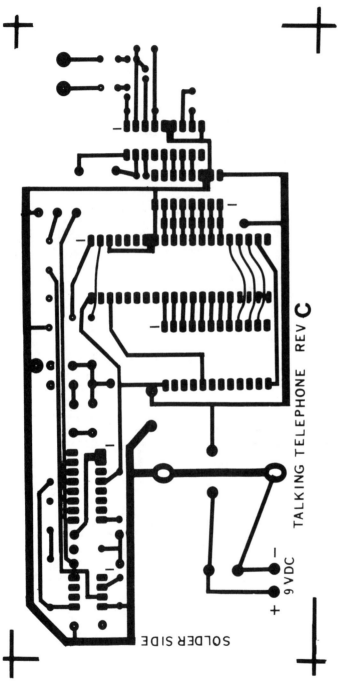

Fig. 16-7 *Artwork for the solder side of the talking telephone PC board.*

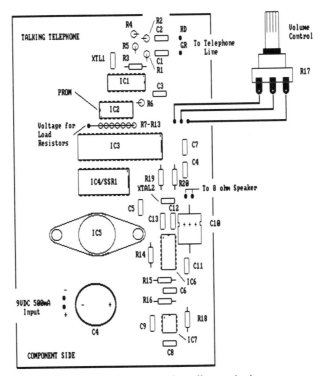

Fig. 16-8 *Parts layout for the talking telephone.*

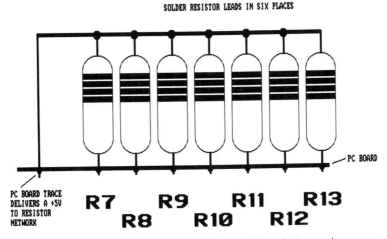

Fig. 16-9 *To save precious real estate, resistors R7 to R13 must be mounted on end. The free ends are soldered together then connected to a pad drilled on the board (+5 V).*

A soldering pencil with a sharp tip is the only way to solder the IC and to prevent shorts. Please be careful when soldering the IC. Do not overheat the device or add solder bridges.

Because the talking telephone uses a double-sided printed circuit board, IC sockets cannot be used. Soldering the underside of the board while using sockets is no problem. It is soldering the upper half of the board that causes the problem. Because IC sockets are made of molded plastic, bringing a hot soldering iron in close proximity to the component will melt it. Melting causes the connector pins to fall out. So while directly soldering the Digitalker and the SSR1 ROM chips into place, be fast, accurate, and by all means, use an alligator clip as a heat sink. The clip will absorb the excess heat from the iron and prevent expensive IC burnouts.

The final items to be mounted and soldered are the speaker leads, volume control, telephone line cord, and the 9 Vdc (500 mA) wall transformer. By observing their location and polarities on the PC board component layout (Fig. 16-8), the installation of these devices should pose no great difficulty.

TESTING AND INSTALLING THE TALKING TELEPHONE

Without connecting the talking telephone to a phone line, just plug in the wall transformer. A female voice will say automatically, "This is Digitalker." This message indicates that a majority of the circuitry is operational. To test the circuit even further, plug the line cord into a standard telephone jack. Lift up the receiver of a tone-dial telephone. Press each button and listen for the appropriate response from Digitalker. When all 10 buttons have been tested and are operating correctly, press the * and the # buttons. At this time, you will hear a tone burst from the speaker.

When not in use, turn the talking telephone off. The addition of a toggle or slide switch in one side of the 9 Vdc power line should be sufficient.

To have the talking telephone project and a telephone connected to the same wall jack, you might consider purchasing a telephone TEE adapter. Figure 16-10 illustrates such a device. These connectors can be purchased from many sources.

FINISHING TOUCHES

The talking telephone can be mounted in just about any type of housing. Whether it is a plastic or metal box, drill vent holes for the speaker. Mount the volume control (R17) on the front panel and top it off with a fancy aluminum knob. To give you that store-bought appearance, Fig. 16-11 provides you with a professional-looking template. Just make a copy of the page on a copy machine. Cut it out and glue it to the front panel of the housing.

However you wish to finish up your project, the talking telephone will provide you with many years of trouble-free operation.

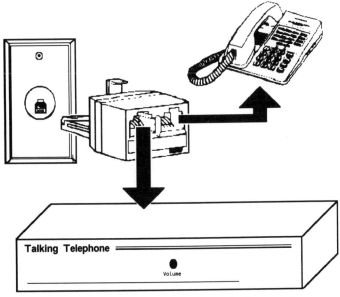

Fig. 16-10 *For easy installation of your finished talking telephone project, use a TEE adapter.*

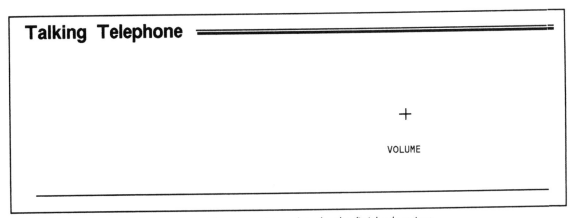

Fig. 16-11 *A front-panel template for the finished project.*

PARTS LIST FOR THE TALKING TELEPHONE

R1 R2 R5	100 kΩ resistors
R3	47 kΩ resistor
R4	49.9 kΩ 1 percent resistor
R6	220 kΩ resistor
R7 to R13 R15	1 kΩ resistors
R14	620 kΩ resistor
R16	10 kΩ resistor

R17	50 kΩ potentiometer
R18	10 Ω resistor
R19	1.5 kΩ resistor
R20	1 MΩ resistor
C1 C2 C6 C11	0.01 μF disk capacitor
C3 C5 C7 C8 C9	0.1 μF disk capacitor
C4	2200 μF 35 V radial electrolytic capacitor
C10	330 μF 35 V axial electrolytic capacitor
C12	20 pF disk capacitor
C13	50 pF disk capacitor
XTAL1	3.58 MHz crystal
XTAL2	4 MHz crystal*
IC1	G8870 DTMF receiver**
IC2	74188 programmed PROM**
IC3	DigiTalker (MM54104)†
IC4	MM52164-SSR1 (part of DigiTalker† IC kit—three pieces) (SSR2 PROM not used)
IC5	7805K 5 V voltage regulator (TO-3)
IC6	LM346
IC7	LM386
T1	9 Vdc, 500 mAdc wall transformer
1	Two-sided printed circuit board**
1	Housing
1	TEE adapter
1	Telephone line cord

---

* This item is available from Jameco Electronics, 1355 Shoreway Road, Belmont, CA 94002.
** This item is available from Del-Phone Industries, P.O. Box 5835, Spring Hill, FL 34606.
†See text.

Synthesized Speech for Your Telephone

THE LAST COUPLE CHAPTERS PRESENTED WAYS TO TRANSFORM YOUR QUIET, nonresponsive telephone into a little chatter box. But even having the phone say words like *attention*, *alert*, *one*, *two*, *three*, and so on, the two circuits are severely restricted by the vocabulary of the Digitalker. The project shown here is not limited solely to telephone circuits but is a building block for more advanced talking designs of the future.

In the next few pages, you will learn how to make a circuit that takes specific words already programmed into the Digitalker's four ROM chips. Then, by linking them together, one right after the other, you can generate phrases or sentences instead of separate words.

If you recall, Digitalker has data and address lines just like a computer. The data lines represent points where the binary equivalent to the word you wish spoken are entered. The address lines select the appropriate speech data from the ROM memory chips (SSR1, SSR2, SSR5, and SSR6). Once the binary-word data is entered (pins 8 to 15), you need to inform the Digitalker that you are ready for an output. This is done by bringing the write strobe (pin 4) to ground.

In addition to the data, address, and the write strobe, Digitalker has another output pin that you will use. This pin is the interrupt. At pin 6, a ground or logic 0 is applied during the normal standby (silence) period. But when you apply a logic 0 to pin 4, the interrupt voltage level increases to logic 1. Then, at the completion of the spoken word, the level will return to logic 0.

You can take advantage of these two pins by applying them to gating circuits that deliver the correct logic levels to the Digitalker at the appropriate times.

Enough talk. See just what it takes to synthesize speech for your telephone.

HOW IT WORKS

The speech circuit is shown in four separate schematics. Do not let the complexity of the design prevent you from building the project. Just take each circuit one at a time and digest its operation. Then collect the parts, and build it. Before long, all four circuits will be brought together into one useful, working project.

Figure 17-1 illustrates the front end of the synthesizer. IC1a/b (4093) is a pulse generator or clocking circuit. This IC delivers a train of pulses (about 2 pulses per second) that are fed into the clock input of (pin 1) of IC2. This 4520 is a dual synchronous divide-by-16 counter (see Fig. 17-5 for a pinout of the IC). With its reset (pin 7) grounded, the enable (pin 2) high, the 4520 will advance a binary count by one on every clock input pulse until the count 1111 is reached. At this point, the IC will reset to binary 0000. This does, however, limit the total number of words in a phrase to fifteen (plus one stop-flag signal). But if you redesign the counter circuit with one that counts to binary 11111, you can easily increase the individual word limit in any programmed phrase to 32.

Whether you use a 16- or 32-bit counter, the data output (D0 to D3 or D0 to D4) is then brought to the input of a programmed 74188 IC where this binary data is used as the address for the desired spoken word. But more on this a bit later.

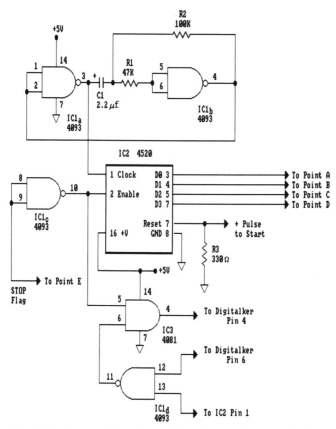

Fig. 17-1 Part 1 of the schematic for the telephone synthesizer
project. This schematic illustrates the clock, counter, and control logic circuitry.

Now, going back to Fig. 17-1, the clock pulses from IC1 are also used to start up the Digitalker (write strobe pin 4) speech circuitry. In addition to starting up the synthesizer, this pulse is also needed at the end of each spoken word to allow a number of words to be strung together into a sentence. Remember the interrupt pin? You can have this logic level going into the input of another NAND gate that is also part of IC1 (IC1c/4093). By applying the clock pulses to one side of the NAND gate and the interrupt signal at the other input, the output will generate one start-up pulse that can be delivered to Digitalker pin 4 at the end of each spoken word. If you use this scheme, you can definitely have the device talk in sentences, but there is no way to stop the circuit. It will just keep on repeating the phrase over and over again. This is where the enable pin of the 4520 counter comes into play. To have the counter count the clock pulses, the enable pin must be at a logic 1. If a logic 0 were placed here at the end of the spoken phrase, IC2 will not accept any more clocking pulses, thus causing the 4520 to stop dead in its tracks. This end of phase pulse (or *flag* as it will be called) can be made part of the 74188 programming logic. Again, more on this a little later.

Still looking at Fig. 17-1, you now have a flag to indicate to the counter that it has reached to end of the phrase, so it stops counting. All well and good. But the output of IC1c (pin 10) will continue generating start-up pulses and delivering them to the Digitalker. With this scenario, Digitalker will say the same word continuously. After a while, the repetition becomes quite annoying. So fix the problem by adding IC3 (4081) to the scheme of things (see Fig. 17-6 for the pinout of the IC). This AND gate will allow an output only when both inputs are high (logic 1). So if you take the NAND output from IC1d (pin 11) and combine it with an inverted stop flag (from pin 2 of IC2) to the inputs of IC3, the start-up pulse will only be generated when pin 5 of IC3 is high and at the logic 1 cycle of the IC1d output. So using this additional circuit, the Digitalker will say what's on its mind, then stop. It will wait to be started up again by placing a positive pulse at the IC2 reset pin (pin 7). By resetting the counter, all data outputs will be forced back to 0000, thus changing the 74188 input address. This change will in turn remove the stop flag signal at the input of IC1c. This will drive the enable pin back to a high level, thus allowing the 4520 to resume its normal counting.

Now turn your attention to the second schematic of the synthesizer (Fig. 17-2). This schematic illustrates the area where, by using two 74188s, you can select the appropriate word you wish Digitalker to say. The binary equivalents for these words are programmed into IC4. IC5 has the pleasure of storing the stop flag pulse at pin 9 and also enabling one of the four Digitalker ROM memory chips. As mentioned in an earlier chapter, both the SSR1 and the SSR2 have the same memory addresses as the SSR5 and the SSR6. If you wish to have the Digitalker say a word that is located within SSR1, you must be able to select this speech data from this chip and not from SSR5. If both ICs were enabled at the same time, unintelligible speech data would be generated. So you must be very selective on where the word that is to be spoken is located. Also that IC must be enabled at the appropriate time and no other. That is the job of IC5.

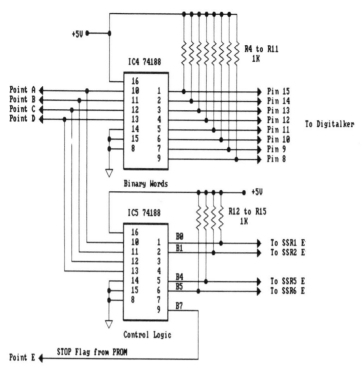

Fig. 17-2   *Part 2 of the schematic for the telephone synthesizer project. Shown here is the wiring for the programmed 74188 ICs.*

Programming of both 74188s will be presented in greater detail a little later in this chapter.

Figure 17-3 shows the heart of the synthesizer. Here the Digitalker IC (IC6) is shown with its four ROM chips. The wiring of this section is a bit confusing because all four memory chips are connected in parallel. In other words, Fig. 17-3 shows that pin 25 of the Digitalker is connected to pin 8 of the SSR1 chip. The schematic also shows that pin 8 of SSR1 is also wired to pin 8 of the SSR2 and to pin 8 of the SSR5 and to pin 8 of the SSR6 memory. This parallel wiring arrangement is also true for memory pin numbers 1 to 7; pins 18, 19, 21, 22, and 23, and pins 9, 10, 11, 13, 14, 15, 16, and 17. See how complicated the wiring can be?

In addition to the read-data and the addresses, the enable pin (pin 20) of each memory chip (SSRx series) must be brought back to its corresponding control pin located at IC5 (Fig. 17-2). To enable any one of the four chips, a logic low (0) must be applied to pin 20 of the appropriate chip.

Figure 17-3 also shows that pin 39 of IC6 is the audio output of the Digitalker and it must be brought to the filter and amplifier section of Fig. 17-4. This filter/amplifier arrangement is the same as used in previous projects, so no great detailed explanation of its operation is given.

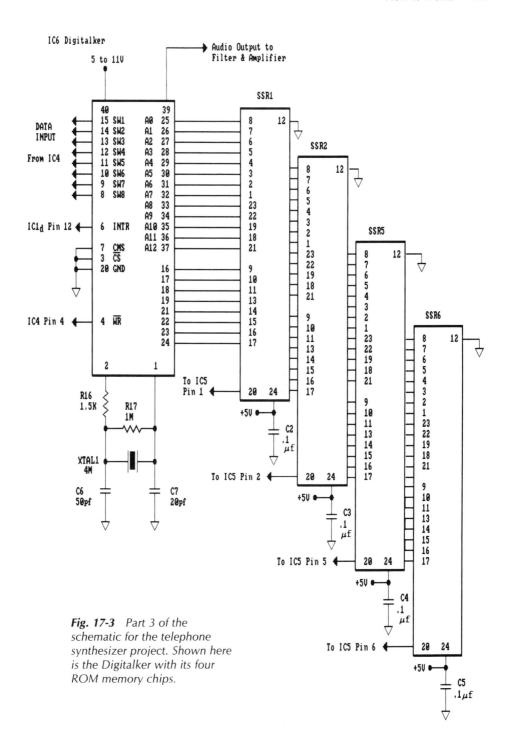

Fig. 17-3 *Part 3 of the schematic for the telephone synthesizer project. Shown here is the Digitalker with its four ROM memory chips.*

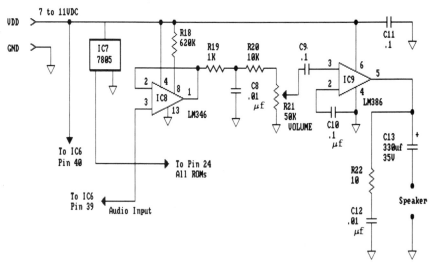

Fig. 17-4   *Part 4 of the schematic for the telephone synthesizer project. Here is the power supply, filter and amplifier circuitry.*

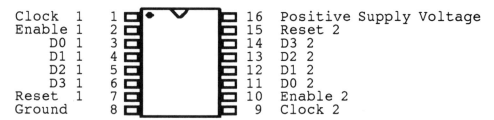

Clock	1	1	16	Positive Supply Voltage
Enable	1	2	15	Reset 2
D0	1	3	14	D3 2
D1	1	4	13	D2 2
D2	1	5	12	D1 2
D3	1	6	11	D0 2
Reset	1	7	10	Enable 2
Ground		8	9	Clock 2

Fig. 17-5   *The pinout diagram for the 4520 Divide by 16 Counter.*

Input	1A	1	14	Positive Supply Voltage
Input	1B	2	13	Input	4A
Output	1C	3	14	Input	4B
Output	2A	4	13	Output	4C
Input	2B	5	12	Output	3C
Input	2A	6	11	Input	3B
Ground		7	10	Input	3A

Fig. 17-6   *The pinout diagram for the 4081 quad two-input AND gate IC.*

You will notice here that the voltage regulator (IC7) is also shown in this illustration. IC7 (7805) is a 5 V voltage regulator IC of the diamond-shaped TO-3 type. Because this IC will be required to deliver enormous amounts of current, it will, without a doubt, create a large amount of heat. This heat, if not drawn away from the regulator, will severely shorten its life expectancy. So to prevent a premature death of the IC, use heatsinks. Presented in Fig. 17-7 are a number of heatsinks that can be used with the TO-3 voltage regulator. These devices are a common component and can be found in just about any electronic store or mail order house for about a buck.

TO-3 Heatsinks

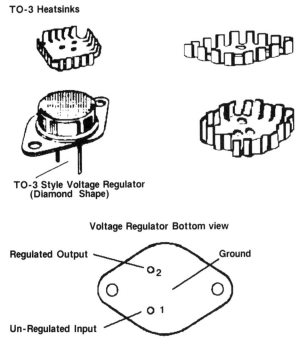

TO-3 Style Voltage Regulator
(Diamond Shape)

Voltage Regulator Bottom view

Regulated Output

Ground

O 2

O 1

Un-Regulated Input

Fig. 17-7   Typical TO-3 heat sinks and pinout diagram.

Like power transistors, voltage regulators have two pins on the bottom. Figure 17-7 shows that pin 1 requires that the unregulated dc voltage be connected here. While pin 2 delivers a healthy ripple free regulated + 5 Vdc output, which is brought to all + 5 Vdc locations on the schematics. The third connection to a TO-3 voltage regulator might not be so obvious. Pin 3 on a TO-3 is the case (or housing) itself. So before wiring the regulator into the circuit, observe the proper pin orientation. Inadvertently applying an unregulated voltage to pin 2 instead of its normal pin 1 location can result in the burnout of the device. Notice how pins 1 and 2 are off centered on the TO-3. This off centering is the only way to tell which pin is which. Keep that in mind during assembly.

ROM ADDRESSING LIMITS

Flip back to chapter 15 and notice that Table 15-2 and 15-3 lists all preprogrammed words that Digitalker is capable of saying. Addresses from 0000 0000 to 1000 1111 can be applied to SSR1 and SSR2 ROMs, and addresses 0000 0000 to 1000 0010 are legal locations for SSR5 and SSR6. Seeing that you must enable the desired ROM chip (through a programmed 74188 IC) when a word located within this ROM is being spoken, you must know the legal-address limits for each memory chip. For this information, see Table 17-1. This table indicates that address locations 0000 0000 ("This is Digitalker") to 0100 0111 (320-millisecond *silence period*) are within the memory range of SSR1. So if you desire to have a word that can be found within this span, you must provide

Table 17-1. The Digitalker memory chips can be divided into four legal address limits.

Legal Address Limits for ROMs			
Word	Binary Address		ROM #
This is Digitalker	0 0 0 0 0 0 0 0		SSR1
320m Second Silence Period	0 1 0 0 0 1 1 1		
Centi	0 1 0 0 1 0 0 0		SSR2
Weight	1 0 0 0 1 1 1 1		
Abort	0 0 0 0 0 0 0 0		SSR5
Key	0 1 0 0 0 0 0 1		
Level	0 1 0 0 0 0 1 0		SSR6
Zone	1 0 0 0 0 0 1 0		

a ground enabling (via a programmed 74188/IC5 memory chip SSR1). Address locations 0100 to 1000 ("centi") to 1000 1111 ("weight") can be found on SSR2. Again, for words located within these memory locations, you must provide a ground enabling the SSR2 memory. To continue, the addresses 0000 0000 ("abort") to 0100 0001 ("key") are all contained within the SSR5 memory. And finally, addresses 0100 0010 ("level") to 1000 0010 ("zone") are located within the SSR6 memory. See how the memory addresses overlay between each pair of ROMs? This is why you need to enable the specific chip that the desired word is located. Otherwise the synthesizer will talk in unintelligible gibberish.

PROGRAMMING THE 74188s

With a large part of the circuit theory behind you, read how you should go about programming the 74188s to generate a synthesized phrase.

Take an easy one to start off with. Program the phrase "Warning, door open, please secure." This phrase is not much good for the telcom industry, but it will demonstrate the flexibility of the circuit. So for now, use Table 17-2 to guide you through the following.

First of all, you have a word *warning*. By going back to chapter 15, you will see that this word has a binary code of 0111 1011. Using one of the PROM programmers presented in appendix B, at address 00, program this binary code into a virgin (new) 74188 chip. Seeing that the next character to be programmed is a comma, which means to pause, take the binary code for a 320 millisecond silence period (see Table 15-2). This 0100 0111 will be programmed into the 74188 at address 01. *Door* is the next word in the sentence

Table 17-2. How to program your first phrase into the 74188 memory chips.

Sentence Programming (IC4 & IC5)

Phrase   ===>   WARNING, DOOR OPEN, PLEASE SECURE <===

Word	ROM #	IC4 Programming B7 ——————— B0	IC5 Programming ROM Enable				
			B0	B1	B4	B5	B7
WARNING	SSR6	0 1 1 1 1 0 1 1	1	1	1	0	0
pause(,)	SSR1	0 1 0 0 0 1 1 1	0	1	1	1	0
DOOR	SSR5	0 0 0 1 1 1 0 1	1	1	0	1	0
OPEN	SSR6	0 1 0 1 0 1 0 0	1	1	1	0	0
pause(,)	SSR1	0 1 0 0 0 1 1 1	0	1	1	1	0
PLEASE	SSR2	0 1 1 1 1 0 0 0	1	0	1	1	0
SECURE	SSR6	0 1 1 0 0 1 1 0	1	1	1	0	0
		END OF SENTENCE "FLAG"	1	1	1	1	1

Logic 1 "Flag" MUST be placed AFTER last word in sentence or phrase.

and its code is 0001 1101. Program this at address 02. The next word is *open*; the code of which is 0101 0100. Program this code at address 03. Again, you have encountered another comma or pause. So use the 320 millisecond silence period for a second time. The binary code for the silence period 0100 0111. Program this silence period at address 04. The next word is *please*. Binary code for this word is 0111 1000. Program this at address 05. The last word is *secure*. The binary code is 0110 0110, so program this at address 06. This completes the programming of the binary words (IC4). Now turn your attention to programming the control logic (IC5) chip.

The word *warning* is located within the realm of SSR6. So you must program the enable output from B5 (SSR6 enable pin) to be grounded with all other enables at a logic 1. So with a new chip in the PROM programmer, flip to address location 00 (Note this is the same address location used to program the word *warning* on the IC4 chip). (The same address locations must be used throughout this programming.) Now at B0 (SSR1 enable), B1 (SSR2 enable), and B4 (SSR5 enable), program a logic 1 at these locations making sure that B5 (SSR6 enable) is left at a logic 0 level. So when the 4520 counter applies the code for the decimal 00, this binary equivalent is delivered to both 74188 ICs. IC4 will develop the coding needed to produce the word *warning*, and IC5 will provide a ground to SSR6 (the memory chip where this speech data is located). The 320 millisecond pause is located on SSR1. So at address 01, bring B1, B4, and B5 to logic 1 while B0 (SSR1 enable) remains grounded.

Door can be found on SSR5. So using the programmer, bring B0, B1, and B5 to a logic 1 at address 02. B4 (SSR5 enable) remains low. *Open* can be found on SSR6. At address 03, bring B0, B1, and B4 to logic 1. This time, B5 (SSR6 enable) remains low.

The next word is the second 320 millisecond silence period, so bring B1, B4, and B5 to a high at address 04. B0 (SSR1 enable) stays low.

The word *please* can be found at SSR2. So at address 05, bring B0, B4, and B5 to a logic 1. B1 (SSR2 enable) stays low.

Finally, the word *secure* can be found on SSR6. With your PROM programmer at address 06, bring B0, B1, and B4 to a logic 1. B5 (SSR6 enable) stays low.

Do not put your programmer away, not just yet. You still need your stop flag signal to be burned in. With the programmer at address 07, bring B0, B1, B4, and B5 and this time bring the flag pin (B7) to a logic 1.

Well, that completes the programming. It was one heck of a job. But you will soon be pleased with the results of your labors.

Take both ICs (hope you know which is which) and plug the chip that contains the binary words into a 16-pin DIP socket at location IC4. Place the other 74188 in the second socket marked IC5. With power delivered to the circuit, the Digitalker should immediately kick into action. The circuit will talk. When the counter has reached the address 07, the flag signal that you have just programmed into IC5, will be applied to the counter enable pin, thus stopping its advancement. There is now silence. To restart the vocal output, just apply a quick positive pulse to the reset pin (pin 7) of the 4520 IC. This pulse will drive all data lines back to 0000, thus removing the stop flag from the counter enable pin. The count and the vocal rendition will resume until, once again, the stop flag is encountered.

Table 17-3 presents a phrase of a different color. Because the vocabulary of the Digitalker is limited to the data stored on the four ROM chips, the total number of words can be extended somewhat by using *homonyms*. Homonym is a term used to describe words that are pronounced the same but have different meanings. An example of this is *ate* (past term of eat, to consume food) and *eight* (the number after seven). Both words are pronounced the same but their meanings are completely different. If you apply this feature to your Digitalker circuit, you can come up with a few additional words for the synthesizer.

Table 17-3. The use of homonyms can increase the usable
vocabulary for the Digitalker. Shown are two homonym words that can be programmed.

Sentence Programming (IC4 & IC5)

Phrase =====> "THANK YOU FOR GOING SLOWER" <=====

Word	ROM #	IC4 Programming B7 ———— B0	IC5 Programming ROM Enable B0	B1	B4	B5	B7
THANK	SSR6	0 1 1 1 0 1 0 1	1	1	1	0	0
U	SSR1	0 0 1 1 0 1 0 0	0	1	1	1	0
4	SSR1	0 0 0 0 0 1 0 0	0	1	1	1	0
GOING	SSR5	0 0 1 1 0 1 1 1	1	1	0	1	0
SLOWER	SSR6	0 1 1 0 1 1 0 0	1	1	1	0	0
		END OF SENTENCE "FLAG"	1	1	1	1	1

**Logic 1 "Flag" MUST be placed AFTER last
word in sentence or phrase.**

Table 17-3 illustrates a phrase that makes use of homonyms. For an example, use the sentence "Thank you for going slower." Because the master word list for both pairs of memory chips does not contain the words *you* and *for*, make use of their homonyms. For the word *you*, the letter *U* (which is available within the SSR1 chip) can be substituted. The number 4 can be a substitute for the word *for* (also available on the SSR1 memory). The meaning of the substituted words in our example are completely different from the intended *you* and *for*, but you are not looking at the meanings of any word but only how it sounds.

To program a phrase such as the one shown in Table 17-3, it does not take any special techniques but rather the same programming procedure presented with Table 17-3. With this phrase, just substitute the binary code of the homonym word for that of the needed sound. Just remember that at the end of every programmed sentence or phrase, you must include the stop flag.

CONSTRUCTING THE SYNTHESIZER

As mentioned before, the speech synthesizer is the most complicated circuit presented in this book. But if you take your time and build the circuit in sections rather than looking at it as a whole, you will be less likely to be intimidated by its complexities.

A circuit such as this is a building block. In all likelihood, the synthesizer will become an intricate part of a more complex circuit design. For this reason, a printed circuit board was not included in this chapter. But even if you construct this device as a stand-alone demonstration unit, using a PCB will not be a cost effective alternative to the inexpensive perf board. To assemble the synthesizer on a PC board, many hours of work would have to be spent designing the layout, making the artwork, then finally etching the board. The time consumed in the fabrication of the final layout can be spent using the old fashioned point-to-point wiring method.

By using the standard 0.1-inch spacing perf board, ICs, everyday components such as diodes, capacitors, and resistors, can easily be mounted. If the need arises, components can be relocated to another area of the board without too much difficulty. Figure 17-8 shows a perf board with 0.1-inch spacing

Fig. 17-8 Perforated (or perf) boards can save you a lot of time in the designing, layout, and etching of a printed circuit board. In some designs, the perf board is the only logical way to go.

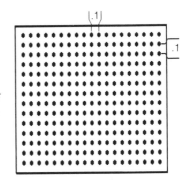

between holes. The physical size of the board can vary from store to store but for the synthesizer, a board no less than 6 inches by 6 inches should be used. Use a larger board, if one is readily available.

With a large perf board on hand, it is time to begin mounting the components associated with the schematic shown in Fig. 17-1. Just in this small circuit, four ICs are used. If you wish, use a DIP socket for each. It is not mandatory, but it would simplify troubleshooting or the replacement of a defective part.

In addition to the ICs, which are polarity-sensitive devices, capacitor C1 must be installed in the circuit with the correct orientation. This means that the positive marking of the capacitor must be connected to the pin 3 output of IC1a (4093 IC). Other than that, Fig. 17-1 should pose no wiring problems.

Figure 17-2 shows the wiring arrangement for the two programmed 74188 ICs. The only problem you will encounter with this circuit is the repetition of the wiring. To help prevent wiring errors, you might consider purchasing IC markers (see Figs. 15-7 and 15-8). These plastic cutouts are numbered in such a way as to correspond with the pinout of an integrated circuit as viewed from the bottom side. They do help in this type of *rats nest* wiring, and they are well worth the added expense.

As with the redundant wiring found with the 74188 memory chips, Fig. 17-3 is just as bad, if not worse. As previously discussed at the beginning of this chapter, all four of the Digitalkers ROM chips are connected in parallel (for example all pin 8s are wired together, all pin 9s are wired together, and so on) with the parallel wiring of 21 separate connections, the operation can become tiresome after a while. So to prevent mis-wires, when you become tired, stop the assembly and go for a walk. Come back a little later. The synthesizer will still be here when you return.

You might have noticed that the Digitalker requires an unusual type for crystal. This 4 MHz device should be acquired along with the Digitalker and its ROM chips from Jameco Electronics. Usually, hobbyist crystals come in two varieties: the large HC33/U and the smaller HC18/U (see Fig. 17-9). Except for their physical size, there is no difference in the oscillator stability or drifting.

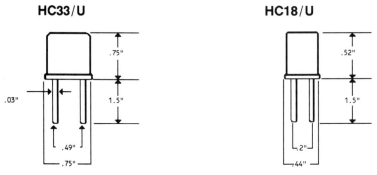

Fig. 17-9 *Crystals come in a wide variety of styles.*
For the hobbyists, the HC33/U and the HC18/U are the most common.

The only distinction between the two is that the HC18/U might be less expensive than its counterpart (HC33/U). You might find a sale on surplus electronic parts, and the 4 MHz crystal can be picked up for a song. So it might be worth your while to investigate the classified ads that can be found in the back of magazines like *Modern Electronics* or *Radio-Electronics*. You will find some pretty exotic parts being advertised there.

The final assembly to be added is the power supply circuit, consisting of the 7805 (IC7/TO-3 regulator) and the filter/amplifier combination. The circuit itself is straightforward, and it should pose no problems in its assembly. Because the completed synthesizer can be considered a futuristic circuit, why not dress up the finished project by substituting a slide potentiometer for the old-fashioned rotary volume control? Figure 17-10 illustrates a typical slide potentiometer that can be purchased from any mail order house for about $2.00. By substituting a slide for the R21 volume control in the amplifier circuit, you can give the device a more pleasing appearance.

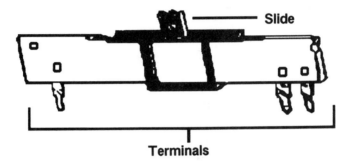

Fig. 17-10 *Giving your project that futuristic style?*
Do not overlook the possibilities that a slide potentiometer can offer.

Again, as with all ICs as well as voltage regulators, proper orientation is the key to successful kit building. So pay close attention to the LM346 and the LM386. IC7 can become a problem in itself. Take special care when wiring the voltage regulator. Remember Fig. 17-7? The only way the proper pin selection can be made is by noting that pins 1 and 2 of the regulator are off centered. With the TO-3 case turned upside down, compare it to Fig. 17-7. If you do, you should not have any difficulties with its connections.

TESTING THE CIRCUIT

The time has finally come. You have soldered that last wire in place. Is it time to install the ICs and plug in the dc wall transformer? Hold on a second. The Digitalker ICs set you back about $27.00 (not including shipping). Do not be in a hurry to plug them in without first visually inspecting the board for miswires, solder splashes, or bridges. A $27.00 investment can be ruined in the blink of an eye. So let's first eyeball the wiring. Then take continuity checks of the wiring for the four ROM memory chips. This is just the place where those little wiring gremlins love to lurk. So get them out of there before power is applied.

When you are sure all the assembly gremlins are eliminated from the circuit, you can now apply power. Do not put in any ICs, not just yet. Power is applied at this stage of the game to check out the voltage levels at all the + 5 V locations. To check these levels properly, you should take the black lead of a VTVM (vacuum tube voltmeter), VOM, or DVM and connect it—not to the common ground point as you might expect—but directly to the grounded pin at the IC socket under test. The red lead can then be placed at the terminal that required the + 5 V (or whatever voltage is required—check schematic). A procedure such as this is used simply because if the black lead from the meter were placed at the common power supply connection and the red lead were placed at various termination points, you will read the normal voltage. OK, so what? Think about this. What would happen if you forget to connect one or more ICs to the common ground? You would still get the normal voltage measurement. But in the real world, there would be no voltage across the IC itself. The IC would be dead. This oversight would play havoc when troubleshooting time comes. So take the advice and check for voltages between the required pins.

All right. You have checked for mis-wires, bridges, voltage levels and continuity of the ROM chips (SSR1 to SSR6). All seems OK. So carefully install the ICs (remember to use a grounding strap) in their respective sockets. Then apply power.

As soon as power is surging through the copper wires of the circuit, the Digitalker will immediately say the phrase or whatever you have programmed into the 74188 (IC5). As soon as the synthesizer reaches the end of the sentence, the circuit should automatically turn off and stay in a standby mode, where it will stay until you place a positive-going pulse on the reset pin (pin 7) of the divide-by-16 counter. That is all there is to it. If you have any difficulty in the operation of the circuit, you should use common troubleshooting procedures to pinpoint the problem and professional soldering methods to correct the difficulty.

PARTS LIST FOR THE SYNTHESIZER

R1	47 kΩ resistor
R2	100 kΩ resistor
R3	330 Ω resistor
R4 to R15, R19	1 kΩ resistors
R16	1.5 kΩ resistor
R17	1 MΩ resistor
R18	620 kΩ resistor
R20	10 kΩ resistor
R21	50 kΩ potentiometer (or slide potentiometer)
R22	10 Ω resistor

C1	2.2 μF, 16 V electrolytic capacitor
C2 to C5, C9,	
C10, C11	0.1 μF disk capacitor
C6	450 pF disk capacitor
C7	20 pF disk capacitor
C8, C12	0.01 μF disk capacitor
C13	330 μF 35 V electrolytic capacitor
IC1	4093 2-input NAND Schmitt trigger IC
IC2	4520 divide-by-16 counter
IC3	4081 2-input AND gate IC
IC4	74188 programmed with desired words
IC5	74188 programmed with control logic
IC6	Digitalker IC*
SSR1 SSR2 SSR5 SSR6	Digitalker ROM memory chips*
IC7	7805 5 V voltage regulator (TO-3 Case)
IC8	LM346 operational amplifier IC
IC9	LM386 2 W mono amplifier
Speaker	8 Ω speaker
XTAL1	4 MHz crystal*
1	12 Vdc 500 mA wall transformer
1	Housing

*This item is available from Jameco Electronics, 1355 Shoreway Road, Belmont, CA 94002.

☎A
Telcom Cookbook

USING RELATIVELY SIMPLE ELECTRONIC COMPONENTS SUCH AS RESISTORS, CAPACI-
tors, and easily obtainable ICs, a number of sophisticated projects have been
constructed. These enhance the performance of a common, everyday telephone
receiver.

Here are cookbook circuits that can be easily added to your own telephone
designs. The schematics are not complete circuits, but with the addition of a
few more components and a little bit of imagination on your part, you can magically transform them into rather elaborate, high-tech telephone enhance-
ments.

POLARITY GUARD

Figure A-1 illustrates a circuit that will allow a tone-dial telephone to be
connected to a line without worrying if the polarity is correct. All that is necessary is to add a bridge circuit (represented by diodes D1 to D4) inside the telephone instrument itself. Note that the connecting wires for the telephone
ringer are connected to the line before the bridge circuit.

PULSE SUPRESSORS

Figure A-2 corrects a sticky situation that you might have encountered
while adding an enhancement device to a rotary telephone. The problem is the
generation of high-voltage pulses during the dialing.

By placing this simple circuit in parallel with the existing telephone line,
the pulses are suppressed by the forward and reverse biasing of the two diodes
and the filtering action of the 0.47 μF capacitor.

A circuit of this type can be used at the input to a telephone ring detector
or similar device. For an electronic ringer, the dialing of a rotary telephone will
not produce *spikes* that can be incorrectly interpreted as a ringing signal and
provide false signals to the circuit.

To further eliminate false detection of the ring signal, Fig. A-3 illustrates
yet another circuit that prevents spurious pulse generation. By applying the
output of a ring detector and a voltage generated by a charging capacitor
(10 μF) to the input pins of a NAND gate, you can literally filter out any
unwanted dialing spikes that can produce erratic operation in any telephone
project.

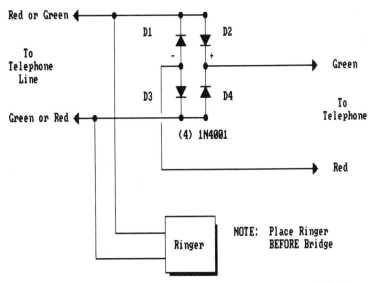

Fig. A-1 *A polarity guard addition for just about any type of telephone.*

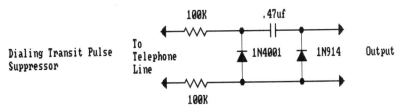

Fig. A-2 *A circuit to suppress dialing transit pulses on a rotary telephone.*

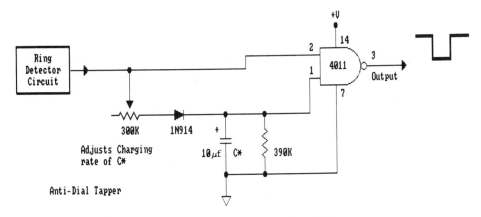

Fig. A-3 *A prototype circuit that eliminates
the tapping of a telephone's ringer when dialing a number on a rotary phone.*

COMMON-GROUND CONVERTER

Normal electronic devices, such as amplifiers, have one of their inputs connected to a common grounding point such as the chassis. In the world of telecommunications, grounding one side of the telephone line is an FCC no-no. These lines are balanced, and they must stay balanced. The only way around this situation is to incorporate a circuit known as a differential amplifier. Figure A-4 shows such a circuit. By using an operational amplifier, the needed ground reference point for an IC can be generated while maintaining the balance of the telephone line. Note that a split power supply is needed in this circuit.

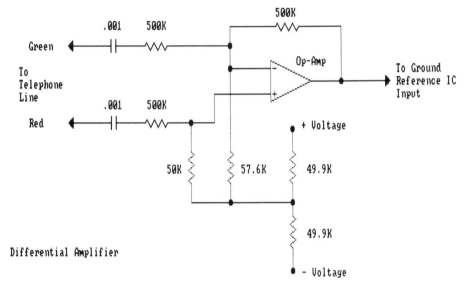

Fig. A-4 A front-end circuit used to convert
the standard telephone line into a common-ground configuration.

ON- AND OFF-HOOK DETECTORS

Many telephone circuits you will design require that they detect either the incoming ring signal or the removal of the handset from its cradle. To distinguish between the on- and off-hook state of the telephone, Fig. A-5 depicts a circuit that does just that. By keeping an electronic eye on the voltage level of the telephone line, you can easily create a voltage to flip on a relay.

By using the internal LED of an opto-isolator IC, you can detect whether or not the handset is on-hook. If the telephone handset is on its cradle, the voltage across pins 1 and 2 of the 4N33 will be zero. But if the handset is removed, the normal talk battery voltage is now placed across these same two pins, thus lighting the internal LED.

If you recall from a number of the projects, the light generated by the LED falls upon a light-sensitive transistor. By applying a voltage at the collector of

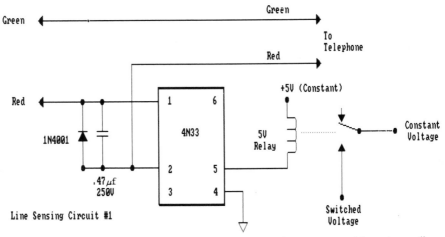

Fig. A-5 *A circuit that senses when a telephone handset is removed from its cradle.*

this transistor through a load resistor (from Fig. A-5, this load is in the form of a relay coil), a voltage drop is created by the flow of electrons through the transistor. This in turn will allow the relay to pull in. The contacts of the relay can be wired to a circuit requiring the detection of a handset off-hook.

Another way to detect whether a handset is off-hook is presented in Fig. A-6. Here, all you do is to install a 5 V *reed relay* in series with one side of the telephone line (red). When the handset is off-hook, a voltage is allowed to flow through its coil, closing the contacts.

A circuit of this type should be used for experimentation only. By wiring the coil of a relay as shown, the telephone line is no longer balanced. To rebalance the line, just measure the resistance of the relay coil and place a resistor of

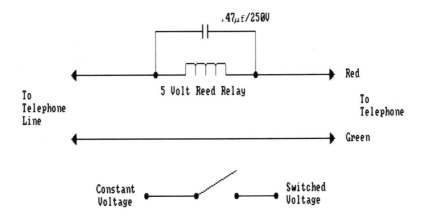

Fig. A-6 *Another line-sensing circuit. In this circuit, a relay is placed in series with the line.*

equal value in series with the second telephone wire (green). But there is a better solution to this balancing problem. Just read on.

The proper balance of a telephone line must always be taken into consideration when designing any type of telcom circuit. FCC rules regarding cus-
tomer-provided equipment must also be considered. So to keep the FCC
appeased, it is a good idea to incorporate devices that already have an FCC
approval in your design. Such a device is the Teltone line-sense relay as pictured in Fig. A-7.

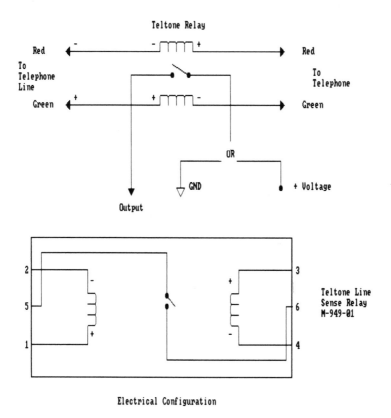

Line Sense Circuit #3

Fig. A-7 *Yet another line-sensing circuit*
that makes use of the FCC approved Teltone M-949-01 relay.

This circuit, like its predecessor (Fig. A-6), detects the on or off status of
the telephone handset but with one big difference. By making use of the
Teltone's M-949-01 relay, a second balancing resistor need not be taken into
consideration. Figure A-7 shows that the M-949 uses two individual relay coils
of identical resistance. When placed in series with the red and the green wires
of the line, these coils provide the needed line balancing.

As with Fig. A-6, the Teltone relay is activated when the handset is lifted from its cradle. The relay drops out when the handset is replaced. The relay contacts can be wired to give either a ground or voltage when closed.

Note that the line-sense relay requires that the correct voltage polarity be applied to its coils for proper operation. Other than that, the M-949-01 works like any other relay.

TONE AND RING DETECTORS

For a telephone to operate correctly, the central office provides signals or tones to let you know the status of a telephone call. This is, the telephone company generates a dial tone when the handset is lifted. This indicates that you can now dial the number of the party you wish to contact. They also provide a tone to let you know that the telephone is ringing at the dialed number. If that number is busy, you hear another type of tone.

If you design a telephone device that connects across the line, how can you have your project distinguish between these tones? You can provide the needed capability by using the specialty ICs shown in Figs. A-8 and A-9. With these tone detectors, your black box can distinguish between the four common signalling tones.

Figure A-10 illustrates the signalling tones in question. The basic tones are: dial tone, ring back, and busy signal. By incorporating the Silicon Systems SS75T982 IC into your next project, detection of these tones are a snap. So do not overlook these two versatile telcom ICs in your next design.

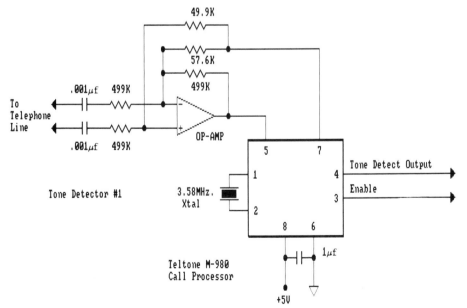

Fig. A-8 *A telephone tone detector using the Teltone M-980 call processor IC.*

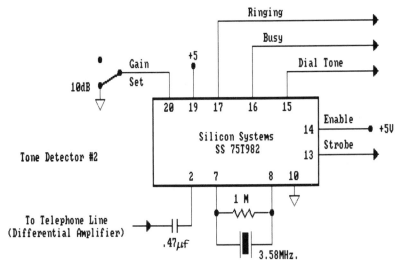

Fig. A-9 *Another telephone tone detector. This circuit uses the Silicon System SS 75T982 IC.*

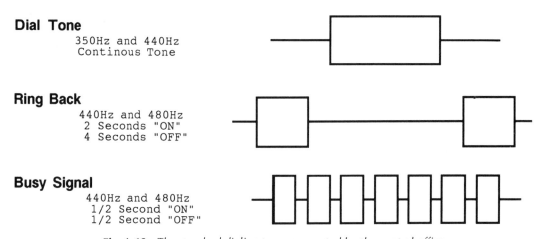

Dial Tone
      350Hz and 440Hz
      Continous Tone

Ring Back
      440Hz and 480Hz
      2 Seconds "ON"
      4 Seconds "OFF"

Busy Signal
      440Hz and 480Hz
      1/2 Second "ON"
      1/2 Second "OFF"

Fig. A-10 *The standard dialing tones generated by the central office.*

Another ring detector design is shown in Fig. A-11. This circuit uses the standard 4N33 opto-isolator IC to provide a logic low in the presence of the ring signal. To provide a snap on/off pulse, the Schmitt trigger NAND gate is used. The 4093 (CMOS) not only provides the snap-action pulses, it also generates an output voltage that is in complement to its input. The output of the gate will be opposite the input.

ZENER OFF-HOOK DETECTOR

Figure A-12 shows yet another off-hook detector circuit. This time a zener diode is used to detect the 50 V on-hook or the 5 V off-hook status of the handset.

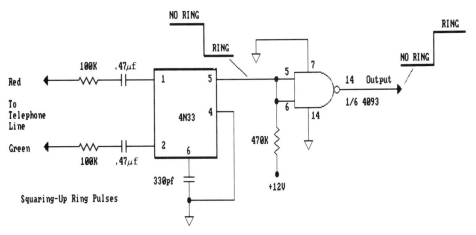

Fig. A-11 *If your design needs a square-up ring signal, this circuit is just what the doctor ordered.*

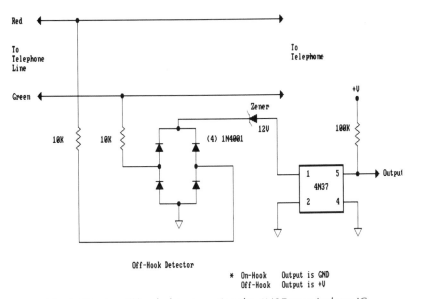

Fig. A-12 *An off-hook detector using the 4N37 opto-isolator IC.*

FREE POWER SUPPLY

Would you like to power your next project using voltage supplied by the telephone company? With the circuits presented in Figs. A-13 and A-14, you can do just that. In both schematics, a zener is used to clamp the telephone line voltage to a 11 Vdc, and the 330 μF 50 V capacitor is used to filter out any minor voltage interruption in the line. The transformers are used to provide a means of impressing or removing an audio signal across the line.

If you wish, you can connect the secondary of the transformer to the input of a 250 mW audio amplifier. The results would be a basic speaker phone with

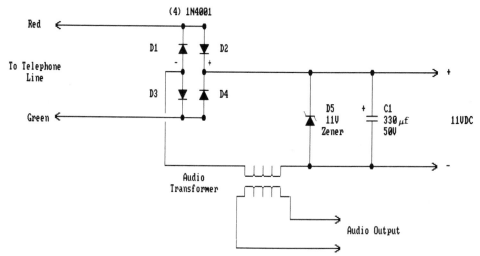

Fig. A-13   Need voltage to power your next telephone project?
How about adding this project and make the telephone company foot the bill?

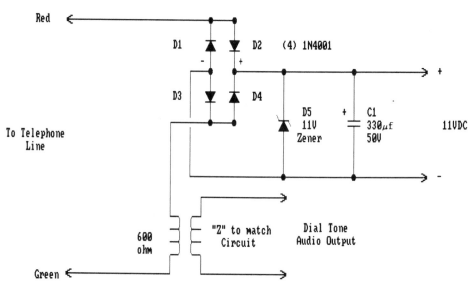

Fig. A-14   Another power-stealing circuit for your next project.

free power. The transformer for the speaker phone can be connected in series
with the negative side of the derived power supply circuit as seen in Fig. A-13.
As an alternative, the transformer can be connected (also in serial) with the
green line cord wire as shown in Fig. A-14.

UNIVERSAL TELEPHONE INTERFACE

Figure A-15 shows a complete telephone interface. IC1 provides a standard
ring detector with its output delivered to pin 5 through a 100 kΩ resistor.

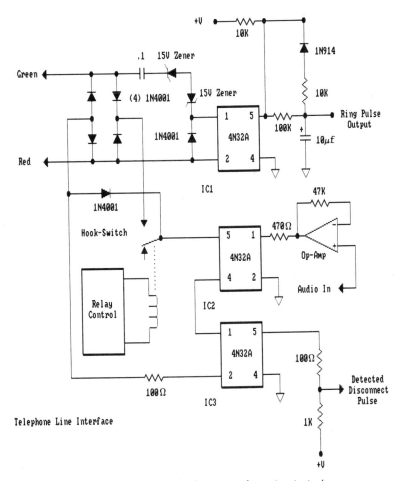

Fig. A-15 *A universal telephone interface circuit. It detects*
a ringing telephone, impresses an audio signal on the line, and detects a disconnect pulse.

IC2 provides a way to impress an audio signal on the telephone line. It makes use of light modulation. Assume an op-amp is connected to the internal LED pins of IC2. With an audio signal applied, the light emitted by the LED will change and be directly proportional to the input. This modulated light is allowed to shine on the light-sensitive surface of the internal transistor of the 4N32A, where the current flow through this transistor is varied by the LED. This variation impresses the signal on the telephone line without the use of a transformer.

IC3 provides a way of detecting a disconnect pulse across the line. Assume that the person you are talking to decides to hang up. This action will generate a short interrupt pulse on the line. With the internal LED of IC3 on during normal conversation, the circuit will sense the break in line voltage by turning the LED off for a short time. This off period will allow a voltage to be present at the junction of the 100 and 1000 Ω resistors.

AUTOMATIC DIALER

The best is saved for last. Figure A-16 is a circuit of an automatic telephone dialer (tone dial). It uses another Teltone IC called the call progress tone generator. By applying a binary code at its input, the M-991 will generate the standard DTMF output. This circuit can actually dial a predetermined telephone number. This circuit can be a start to that home burglar alarm you were meaning to design.

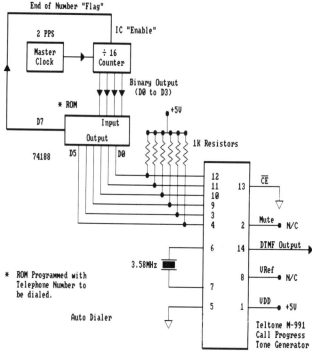

Fig. A-16 By applying a little imagination
and this circuit, you can easily construct an automatic telephone dialer.

But how does it work? Simple. Assume you design a pulse generator (master clock) with a rate of 2 pulses per second. You can then have it connected to the clock input of a divide-by-16 counting circuit. The four-bit binary output of the counter can then be connected to the address-select input pins of a 74188 PROM IC. It is here the telephone number can be programmed.

You can have the 74188 deliver, with each master clock pulse, a predetermined binary code that corresponds to the number you wish to dial. Have it wired to the M-991, where the actual DTMF tones will be generated across the telephone line.

Just remember to program a stop flag into the 74188. The stop flag should be just like the one that was included in chapter 17. This flag will signal that you have reached the end of the programmed number. It should also make the circuit come to a stop until you wish the number to be dialed again.

☎ B

PROM Programmers

A NUMBER OF TELEPHONE PROJECTS IN THIS BOOK MAKE USE OF ELECTRONIC MEMORY. The memory can be added in a number of ways. You might design a circuit using discrete components to achieve the desired binary output. Another way is to connect a number of diodes to a 1-of-16 data distributor IC. This diode matrix circuit can provide, at its output, a nonvolatile, permanent memory cell for your more complex circuit designs. The last memory device to discuss is in a 16-pin DIP IC. Using this miracle of micro miniaturization and a homemade programming device, you can create any binary output from 1 to 8 bits. You can create the output at any one of 32 possible addresses. Compared to the other possibilities, the 8223 or 74188 would be the most economical and easiest to put into practice in our telephone projects.

BITS, BYTES, AND BINARY CODES

For simplicity, Table B-1 will be used as the master listing of the needed binary code that will allow the National Semiconductor's Digitalker to generate a speech pattern. (For an in-depth discussion on the operation of this ingenious integrated circuit, please refer to *The Talking Telephone*.)

Please refer to Table B-1. Note that binary codes (1s and 0s or data bits) are read from right to left. Not the normal left to right. To indicate this reversal to the normal reading direction, many *truth tables* (listings of binary codes generated by digital and memory ICs) contain the letters MSB and LSB. LSB is the least significant bit (First data bit in a series of 4, 8, 16, or even 32 bit-words). MSB indicates that this bit of binary data is the most significant bit of data. The bit indicated as the MSB is the last data bit in a series of 4, 8, 16, or 32 bit words or *bytes* (byte is one complete binary word).

Another method of indicating the same binary information is the use of the letter D (indicates data) and numbers from 0 to 3, 7, 15, or 31. Yes, in the world of binary bits and bytes, 0 (zero) is an active number and place holder, and it should be considered as such. You can combine MSB, LSB and D indicators together and the results can be identified as the following:

$$
\begin{array}{cccc}
\text{MSB} & & & \text{LSB} \\
\text{D3} & \text{D2} & \text{D1} & \text{D0}
\end{array}
$$

Table B-1. A partial word list showing the needed binary codes to command the
National Semiconductor's Digitalker IC to say the words *one* to *nine* and the word *zero*.

Digit Dialed	DTMF IC Output MSB LSB	Required Binary Code MSB LSB	Word Digitalker will say
Power "ON"	0 0 0 0	0 0 0 0 0	This is Digitalker
#1	0 0 0 1	0 0 0 0 1	ONE
#2	0 0 1 0	0 0 0 1 0	TWO
#3	0 0 1 1	0 0 0 1 1	THREE
#4	0 1 0 0	0 0 1 0 0	FOUR
#5	0 1 0 1	0 0 1 0 1	FIVE
#6	0 1 1 0	0 0 1 1 0	SIX
#7	0 1 1 1	0 0 1 1 1	SEVEN
#8	1 0 0 0	0 1 0 0 0	EIGHT
#9	1 0 0 1	0 1 0 0 1	NINE
Operator (0)	1 0 1 0	1 1 1 1 1	ZERO

The least significant bit, as shown above, can also have the name D0. The second bit has the name D1. The third has D2 and the fourth and last (most significant bit) as seen in this example, is named D3. For binary codes using more than four bits, you can extend the D0, D1, D2, (and so on) indicator to the eighth, sixteenth, or even the thirty-second data bit. For the projects contained in this book, you will use only four- and eight-bit data words. You will not encounter 16-bit words.

As for the wiring of a Digitalker circuit, National Semiconductor makes use of the flags (or indicators) SW1 to SW8. SW1 is used to indicate that binary bit D0 should be connected to this point. SW2 signifies that the second binary bit (D1) be connected here. SW3 connects to D2, SW4 connects to D3, and so on until SW8 connects to the binary bit D7.

Whatever scheme is used to show data bits, the lowest number of flag (SW1 or D0) indicates the LSB, and the highest (SW8 or D7—could even be higher if needed) is the MSB.

DIGITAL MEMORY USING DISCRETE COMPONENTS

Now that you know what bits, bytes, truth tables, and binary codes are, put this knowledge to use by designing some memory circuits.

The first memory project is the discrete component memory device. Figure B-1 is an example of this type of memory. This design converts a specific binary code that is placed at its input into another code that is totally foreign in nature to the input.

First take a look at Table B-1. Assume that you want to design a circuit that will generate binary codes that can be input to the data line of the National Semiconductor's Digitalker integrated circuit. If you dial the number 5, you wish to generate the binary code (0101). This code, when applied to the input of Digitalker, will allow the IC to say the word *five*. If you dial the number 9, Digitalker will say *nine* and so on for the other seven numbers. From the time

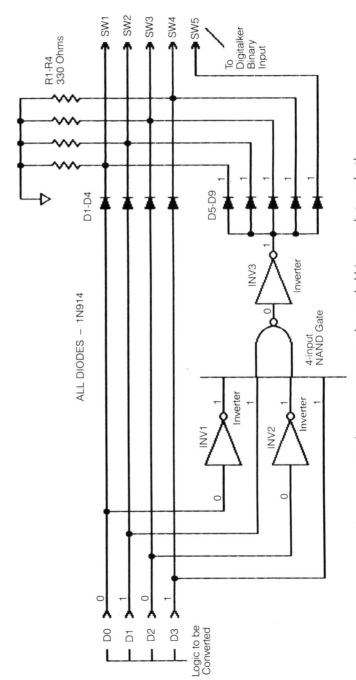

Fig. B-1 *A complex way to generate the needed binary code to make the Digitalker say the word zero. This circuit can be considered to be a simple Memory Device.*

you apply power to the circuit (see Table B-1) to the time we dial the digit 9, the binary code advances in increments of one. Up to this point, you have a standard binary output. There are no significant problems to cope with. Not until you decide to dial the operator (0). Table B-1 shows that the standard binary output of either a Teltone pulse counter or DTMF IC is 1010. By referring to Table B-2, Digitalker partial word list, you can see that this code will make the IC say the word ten. You would rather have Digitalker say the word zero. To produce this word, Table B-2 indicates that the applied binary code to the Digitalker data input (SW1 to SW5, while SW6 to SW8 are grounded) must be 11111. Here you must have some electronic way of taking the normal binary code for the number 10 (1010) and converting it and only it to a secondary binary code made up of five 1s (11111).

Table B-2. A more extensive word list for the Digitalker.

Digitalker Partial Word List

WORD	8-BIT BINARY ADDRESS	WORD	8-BIT BINARY ADDRESS
	SW8 SW1		SW8 SW1
This is Digitalker	0 0 0 0 0 0 0 0	Sixteen	0 0 0 1 0 0 0 0
One	0 0 0 0 0 0 0 1	Seventeen	0 0 0 1 0 0 0 1
Two	0 0 0 0 0 0 1 0	Eighteen	0 0 0 1 0 0 1 0
Three	0 0 0 0 0 0 1 1	Nineteen	0 0 0 1 0 0 1 1
Four	0 0 0 0 0 1 0 0	Twenty	0 0 0 1 0 1 0 0
Five	0 0 0 0 0 1 0 1	Thirty	0 0 0 1 0 1 0 1
Six	0 0 0 0 0 1 1 0	Forty	0 0 0 1 0 1 1 0
Seven	0 0 0 0 0 1 1 1	Fifty	0 0 0 1 0 1 1 1
Eight	0 0 0 0 1 0 0 0	Sixty	0 0 0 1 1 0 0 0
Nine	0 0 0 0 1 0 0 1	Seventy	0 0 0 1 1 0 0 1
Ten	0 0 0 0 1 0 1 0	Eighty	0 0 0 1 1 0 1 0
Eleven	0 0 0 0 1 0 1 1	Ninety	0 0 0 1 1 0 1 1
Twelve	0 0 0 0 1 1 0 0	Hundred	0 0 0 1 1 1 0 0
Thirteen	0 0 0 0 1 1 0 1	Thousand	0 0 0 1 1 1 0 1
Fourteen	0 0 0 0 1 1 1 0	Million	0 0 0 1 1 1 1 0
Fifteen	0 0 0 0 1 1 1 1	Zero	0 0 0 1 1 1 1 1

Figure B-1 shows just that kind of circuit. The standard binary code is applied to the D0, D1, D2, and D3 data inputs. The output binary code can be seen at SW1, SW2, SW3, SW4, and SW5 data outputs. The heart of this binary converter are the three inverters (INV1, INV2, and INV3) and the four-input NAND gate. By applying a little digital electronics, you can design a circuit that sees or detects a predetermined pattern of 1s and 0s. Once the pattern is detected, the circuit does something to indicate that this comparison has been made and is determined to be true. To act as this detector, use the four-input NAND gate ½ SN7420 TTL (transistor-transistor logic) IC. The output of the SN7430 will be at logic 0 only when all of the inputs are at logic 1. Because you have a binary input (D0 to D3) of 1010, you must find a way of converting the logic 0s now located at D0 and D2 into a logic 1. For this, use the SN7404 hex inverter IC. This integrated circuit contains six independent inverters. Each one of these inverters will literally generate a binary bit that is opposite to the applied (or input) bit. In other words, if you connect one of the six 7404s inverters (in Fig. B-1) to data bit D0 and the second inverter to D2, you can convert the logic 0 applied to its input into a logic 1 at the output.

Now if you look at the output of the two inverters and the binary data located at D1 and D3, you have an output of four 1s. These four 1s can now be applied to the input of the four-input NAND gate. With the binary code applied, as seen in Fig. B-1, the output of the NAND gate will be a logic 0 only when the applied binary code represents the number 10. For a telephone dial, it represents the digit "0" (or operator).

Now that you have detected the binary code for operator, you still must convert it into a code representing 11111. At this time, the output of the NAND gate is at logic 0. To convert this 0 logic into a 1, connect the third inverter to the output of the 7430 (NAND gate). With this connection made, we will have a 1 at the output of INV3 only when the applied binary code is 1010.

This logic 1 output (output from INV3) is then applied to five 1N914 diodes (D5 to D9—not to be confused with the data inputs D0 – D3). The logic 1 voltage will be allowed to flow through the diode where it is supplied to SW1 to SW5 of the Digitalker data input. Just remember that the additional data inputs (SW6, SW7, and SW8) needed by Digitalker are considered to be at a logic 0. So to provide this needed logic, just connect these three points to ground.

Diodes D1 to D4 in the standard data bus (D0 to D3) wiring, and diodes D5 to D9 are used to prevent the logic 1 voltages being generated by the detection circuit (NAND gate and inverters) from being applied back into the D0 to D3 data bus, thus causing problems. Resistors R1 to R4 are 300 Ω, 1/4 W components. Their main function is to keep the Digitalker data input at a logic 0 at all times unless a logic 1 is applied at the desired time.

Due to the complexities of this circuit, use this type of memory only in a breadboard or prototype wiring configuration. Use it only if these is a number of unused gates and inverters that can be wired into a conversion circuit.

A circuit of this type should not even be considered for printed board designs. Not only is the circuit presented in Fig. B-1 complex, it will also take additional time and money to lay out the needed PC board artwork for such a circuit. With all your applied efforts, you will come up with a circuit that will convert only one binary code into a predetermined output. This scheme is not cost effective.

For more design flexibility, the second electronic memory device presented here will give you more programming options using a handful of 1N914 diodes.

THE DO-IT-YOURSELFER'S READ-ONLY MEMORY

In the heading of this section, are the words Read-Only memory. A read-only memory (or ROM) is a device that is literally taught a desired truth table. Table B-1 is an example of a truth table. Using a ROM, a manufacturer or you (using a special programmer) can educate a dumb chip into remembering that for every selected binary input, you (or a manufacturer) wants a predetermined binary output. A way of teaching an IC a binary code without a programming device is to build, by hand, a semidiscrete read only memory using a readily available 74154 IC. The 74154 is an IC called a 4-line to 16-line decoder. Using this IC, you now have an output from any one of 16 lines, although only ten are shown in Fig. B-2. With a input code applied to pins 20, 21, 22, and 23, the corresponding output line will go to a logic 0, while the rest remain at a logic 1. For instance, apply the binary code 0000 at data inputs D0 to D3. With this code applied, pin 1, which corresponds to the decimal 0 will go to logic 0. With an input code of 1001, the decimal 9 line (pin 10) will go low. The same will happen if a binary input code between 0000 and 1111 is applied to the 74154 IC.

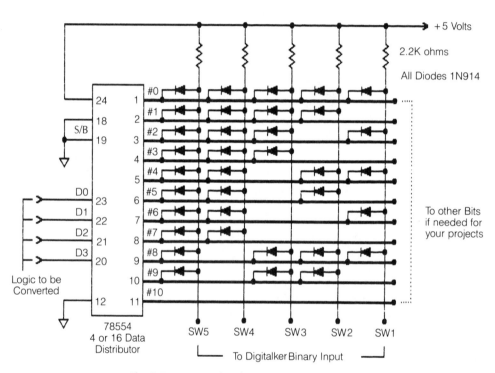

Fig. B-2   A way of making a simple memory
device using the 74154 IC and a handful of diodes and resistors.

Seeing that you need a minimum of five data lines to be connected to the Digitalker, Fig. B-2 illustrates how you can teach the 74154 the needed binary data output. If you compare the diode arrangement of Fig. B-2 to Table B-1, you will notice that for every logic 0 that is needed at the output, a 1N914 diode is placed between the output line of the 74154 and the 2.2 kΩ resistor. The diodes in this scheme perform a negative-logic OR function. This means for every output line of the 74154 that is selected by a binary input and by inserting a diode as shown, the corresponding SWx line will generate a logic 0 while the remaining SWx lines stay high. By adding diodes to every bit location you wish to be low (0), you can reconstruct a large memory circuit that will remember the Table B-1 truth table until diodes are removed from the circuit.

At this time, compare the truth table of Table B-1 with the physical wiring arrangement of Fig. B-2. Notice that when power is first applied, the output of the Teltone DTMF IC will be 0000. To make the Digitalker say "This is Digitalker," a binary input of 00000 must be applied to SW1 – SW5 (data bus for Digitalker). To accomplish this feat, the input code of 0000 is placed at the D0 to D3 data input of the 74154 IC. This input causes pin 1 to go low. By connecting five 1N914 diodes as shown, the diodes shunt the voltage through the now selected 0 line. Thus the output for the applied binary code will be the following:

SW5	SW4	SW3	SW2	SW1
0	0	0	0	0

or

Diode Diode Diode Diode Diode

Now, input the binary code for the number 1. This code will be 0001. Seeing that we wish Digitalker to say the word "ONE" (see Table B-2), the binary output of the ROM can also be the same, except for the added data bit (SW5). To achieve this output, solder four diodes to the SW5, SW4, SW3, and SW2 connection points. The output code will be as follows:

SW5	SW4	SW3	SW2	SW1
0	0	0	0	1

or

Diode Diode Diode Diode OPEN

Note that this scheme can be applied to all the decimal digits up to and including the decimal number 9. When you get to the operator digit, a different code must be generated. Here is how you can do it.

The applied binary input code, generated by a tone or rotary dial IC will be 1010. As stated earlier, if this code were applied to the Digitalker, it will say the word *ten*. But you wish it to say *zero*. So to convert this 1010 to the needed 11111, just leave pin 11 of the 74154 IC open. Do not add any diodes to this line. The absence of diodes will provide a logic 1 to all five outputs. This is what you need to have Digitalker perform the way you want it to.

To aid in the construction of this do-it-yourself ROM, the artwork needed to fabricate the double sided printed circuit board is included. Figure B-3 shows the artwork of the component side of the board, and Fig. B-4 shows the solder side. When soldering the IC and the diodes, remember to solder both sides of the board. Plated-through holes, also can be purchased when the board is made. Plated-through holes are an expense that is not mandatory. Just take a few extra moments and solder both sides of the board. You will find this method to be more economical.

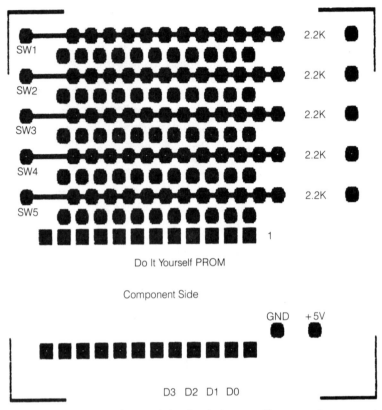

Do It Yourself PROM

Component Side

Fig. B-3 *Computer-generated artwork for the do-it-yourself memory (component side).*

IN REVIEW

Review what you have seen so far. Table B-1 illustrates the binary code needed to generate the correct speech data from the National Semiconductor's Digitalker IC. To accomplish this, the circuit in Fig. B-1 is given. This discrete component binary converter, senses a predetermined binary input and converts it into another, completely different code. The drawback of this circuit is the number of components needed to achieve the desired results. If you require more than two binary conversions, the circuit will be too expensive. Less expensive avenues are open.

Figure B-2 is such an avenue. This do-it-yourself read-only memory can convert up to 16 addresses into any predetermined binary output. Drawbacks of this circuit are:

- A double-sided printed circuit board is needed for operation. Double-sided PC boards are extremely expensive.
- To generate the needed binary output, a large number of diodes are needed. Although the 1N914 diodes are inexpensive, you must consider that you will encounter diodes that will be open, shorted, and leaking.

Components that fail under the temperatures of normal soldering and operation will create unneeded headaches.

- The physical size of the ROM PC board might also prevent you from making use of this technology. The artwork presented in Figs. B-3 and B-4 is shown as twice normal size. When photographed by a PC board manufacturer, this artwork will be reduced in size by 50 percent. Even at this size (2 inches by 2 inches), its use in one of your designs might be prohibited.

So with all this in mind, consider a device that can provide 32 different input binary addresses and provide an output that contains eight bits. Not only will this device supply additional memory locations, it all fits within a body of a standard 16-pin integrated circuit. The component is the 8223 and the 74188 PROM.

You can save valuable board space if you use this IC, but a special piece of equipment must be made in order to teach the chip the required binary truth table. This PROM and programmer are the next topic of conversation.

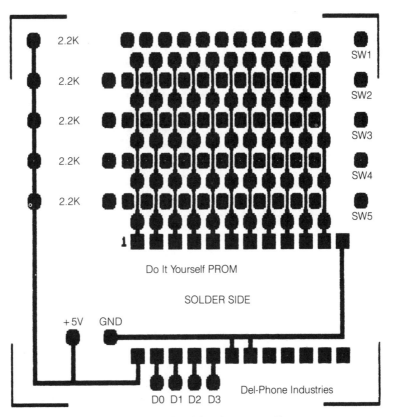

Fig. B-4 *The solder side of the do-it-yourself memory.*
(Both Fig. B-3 and B-4 show the artwork at a scale of 2 to 1.)

MEMORY CHIPS

Due to the tremendous growth in micro miniaturization technology in recent years. Especially in the field of computer memory chips, inexpensive and easily programmed devices have made their way down to the level of the novice and home hobbyist. Quality memory chips can now be found in many major electronic mail order houses as *prime* components and at smaller houses that purchase log-priced Surplus chips.

Basically, memory chips come in two distinct styles. One being the PROM, which you will be interested in, and the EPROM. As mentioned earlier, PROM stands for the programmable read-only memory. Once programmed by a special device, these ICs cannot be reprogrammed. The binary codes burned in to the IC will remain. If other truth tables are needed, the PROM containing the old codes will have to be thrown away. Technology has advanced to the EPROM (erasable programmable read-only memory). This device can actually be erased by allowing an UV (Ultraviolet) light source to shine on the IC. Within 12 to 15 minutes, the original truth table programmed on the EPROM has been erased. The same IC can now be reprogrammed with a new code.

Other memory devices have also hit the electronics market in recent years such as the static RAM, dynamic RAMS, and EEPROMs, electrically erasable programmable real-only memories.

As time goes by, more and more memory ICs are being introduced into the market place especially by the Japanese. The Japanese claim approximately 60 percent of the world market in the advancements in memory technology.

The United States is in the midst of developing an IC that can actually develop its own internal structure. This technology makes extensive use of highly intelligent computers. The thought of having electronic hobbyists, in the near future, sitting down at their computers and fabricating ICs to their own specifications might sound like a fantasy now, but it can become the reality tomorrow.

THE PROGRAMMABLE READ-ONLY MEMORY (8823/74188)

Figure B-5 illustrates the memory IC that you will make extensive use with many of the telephone projects. The 74188 (or 8223) IC memory is contained in the familiar 16-pin DIP. While thumbing through magazines like *Popular Electronics*, old *Hands-On-Electronics* magazines, *Radio-Electronics*, and so on, take note of the advertisements usually found near the back. These advertisements contain a wealth of information and pricing on any type of electronic gear you might need. Look under the category of PROMS, and make note of the pricing of the 74188 IC. For only about $1.50, you can purchase an IC that can be taught to remember 32 words (addresses). Each word can contain eight bits of information. By comparing this IC to the previously discussed memory devices, the 74188 is the most economical to operate and would take up the least amount of board space.

For the proper operation of the 74188, each output pin (B0, B1, B2, and so on) that is being used in your design must have a 1000 Ω, resistor connected to

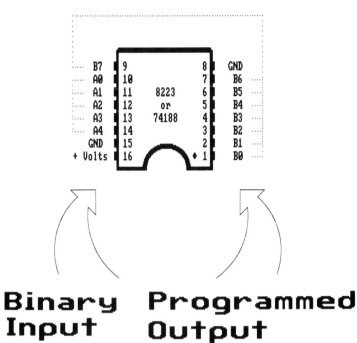

Fig. B-5 The pinout of an electronic memory device, the 74188 (or 8223) IC.

it. The other side, of which is connected to the +5 V power supply. This scheme is needed because the 74188 is manufactured as an *open-collector* device. Figure B-6 illustrates what is meant by an open collector. Picture the 74188 as a transistor circuit. If a transistor is wired using a power supply with a negative ground, the emitter leg is soldered to the negative voltage, and the collector must be connected to the positive side. But as you can see, the output

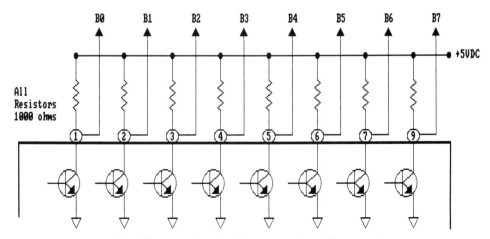

Fig. B-6 A simplified diagram showing the output pins of the 74188/8223 PROM IC.

pins of the 74188 are not connected to anything. The resistors usually associated with this type of transistor configuration have been omitted and must be
replaced using standard carbon-compound devices. The output for each leg is
taken directly from its associated pin. By connecting a voltmeter to these
points, you will see a positive 5 V.

HOW THE 74188 IS PROGRAMMED

The easiest way to explain how the 74188 is programmed is to think of the
IC as the electrical wiring in your home or apartment. Electrical current is
delivered to your home from the power company. Your house wiring brings
this electron flow into the fuse box where a fuse is placed in series with the
power line. Then from the fuse, a line is brought to the ac outlet on the wall,
where a TV is plugged in (see Fig. B-7).

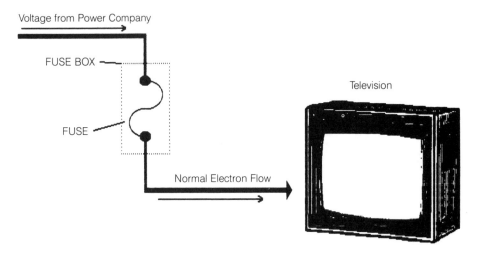

Fig. B-7   *This diagram illustrates the normal electron
flow from the power company, through the fuse, then to a TV set.*

Using this scheme, if the TV were turned on, an electrical flow will be provided from the power company line through the fuse to power the TV. If you
also had the toaster, microwave, and coffee pot connected to the same line, (see
Fig. B-8) there would be excessive electron flow. This excessive flow would
heat up the house wiring, and the fuse would melt. In the terminology of electronics, the fuse will be *blown*. This blown fuse would prevent any flow of
electrons in the line. The line would be considered to be dead.

Programming the 74188 can be considered to be the same principle. The
normal voltage applied to pin 16 of the IC is + 5 V. If pin 16 were to have a + 12
V applied, and at the same time another + 12 V were delivered to an output pin
(B0, B1, B2, and so on), the internal fuse of the IC would be blown. This blown
fuse now corresponds to a logic 1 output. If a logic 0 were needed, the internal
fuse of that particular output pin would not be blown.

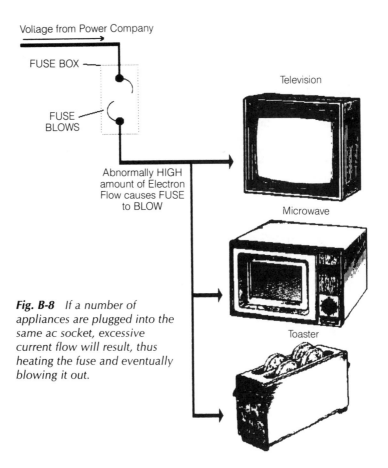

Voltage from Power Company

FUSE BOX

FUSE
BLOWS

Abnormally HIGH
amount of Electron
Flow causes FUSE
to BLOW

Television

Microwave

Toaster

Fig. B-8 *If a number of
appliances are plugged into the
same ac socket, excessive
current flow will result, thus
heating the fuse and eventually
blowing it out.*

Just like the example of the power line and TV (Figs. B-7 and B-8), if the normal voltage used to drive an appliance were increased, the associated fuse will blow. But in the case of the TV set, the blown fuse can be replaced and the TV will operate again. In the case of the 74188, fuses cannot be replaced. So once a logic 1 is programmed, it is programmed for keeps. You can change a pre-programmed truth table somewhat by changing any logic 0 output into a logic 1, but you cannot change a 1 into a 0. If you make a mistake, the IC is ruined. Be happy that the price of the IC is so low.

WHAT IS A DATA ADDRESS?

Now that you can program your truth table into an IC, you need a way of addressing different locations within the IC. But first, an explanation of an *address.* The word *address,* is just what it means. The address is a specific location within memory, just like the address of your home or apartment. Say your address is 97-14 223 Street. To the post office, your home is located at the intersection of 97th Avenue and 223rd Street at the 14th house on the block. Memory chips need the same information. But instead of using the designations avenue and streets, use the standard binary code that will represent decimal numbers.

Table B-3 lists all the binary codes needed to address 32 possible locations within the 74188. You can tell the IC that you want a code of 01101110 to be stored at location (or address) 10000 binary (or location 16 decimal).

Table B-3. This table shows the decimal-to-binary equivalents for the numbers 0 to 31.

Decimal	Binary					Decimal	Binary				
	SW7				SW3		SW7				SW3
0	0	0	0	0	0	17	1	0	0	0	1
1	0	0	0	0	1	18	1	0	0	1	0
2	0	0	0	1	0	19	1	0	0	1	1
3	0	0	0	1	1	20	1	0	1	0	0
4	0	0	1	0	0	21	1	0	1	0	1
5	0	0	1	0	1	22	1	0	1	1	0
6	0	0	1	1	0	23	1	0	1	1	1
7	0	0	1	1	1	24	1	1	0	0	0
8	0	1	0	0	0	25	1	1	0	0	1
9	0	1	0	0	1	26	1	1	0	1	0
10	0	1	0	1	0	27	1	1	0	1	1
11	0	1	0	1	1	28	1	1	1	0	0
12	0	1	1	0	0	29	1	1	1	0	1
13	0	1	1	0	1	30	1	1	1	1	0
14	0	1	1	1	0	31	1	1	1	1	1
15	0	1	1	1	1						
16	1	0	0	0	0						

Figure B-9 shows how you can address the 74188. By flipping SW3, SW4, SW5, SW6, and SW7 to the proper sequence (leaving a switch up will provide a logic 0 to its associated pin, and closing the switch will provide a logic 1), you can select decimal addresses.

For an example, assume you wish to address five sequential locations of the IC. SW3 to SW7 would be flipped to the following sequences:

Decimal Number	Address in Binary				
	SW7	SW6	SW5	SW4	SW3
0	OPEN	OPEN	OPEN	OPEN	OPEN
1	OPEN	OPEN	OPEN	OPEN	CLOSED
2	OPEN	OPEN	OPEN	CLOSED	OPEN
3	OPEN	OPEN	CLOSED	OPEN	OPEN
4	OPEN	OPEN	CLOSED	OPEN	OPEN

the sequence above decimal 4 is omitted here.

Using the binary code as listed in Table B-3, you can address any IC memory location and its programmed binary code by just arranging five switches into any binary code ranging from decimal 0 (00000) to decimal 31 (11111).

PROGRAMMING THE 74188/8223

Figure B-9 is a schematic of a device that is used to program the desired binary code (or truth table) into the IC. It is a very simple circuit to build and use. All that is needed is a handful of parts and a little time; you are on your way to designing projects with minds of their own.

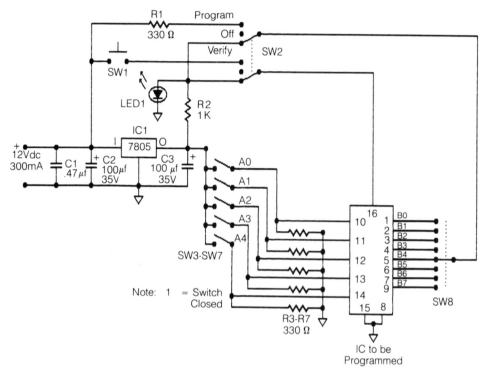

Fig. B-9   *To program a PROM, you need a programmer. This is a simple yet effective way to provide a programmer.*

You first need a dc power supply that is capable of delivering a + 12 V at a current of 300 to 350 mA. This high current can be obtained from a car battery or even a bench-type power supply. Do not try to use a power supply with less than the recommended current rating. This high current is needed to blow the internal fuses of the IC. Remember the power company/television example? Excess electron flow was needed to blow the fuse. You also need excessive electron flow to blow the fuses in the 74188 now.

When you have a power supply of the correct current rating, capacitors C1 and C2 provide a little extra filtering of the voltage. Once filtered, the + 12 V is delivered to a voltage regulator, where it is converted to the normal + 5 V needed by the IC. The 5 V is branched into two locations. One location is the common tie point for the switches used to address the IC, and the second delivers the needed voltage to the 1000 Ω resistor where there is an LED to indicate if a logic 1 or logic 0 is present at the selected IC output.

Switch SW2 is used to switch pin 16 of the IC and the selected output data pins from the standard + 5 to the + 12 V needed to blow the internal fuses of the IC. Resistor R1 (330 Ω) is used to limit the current of the 12 V to a safe level. This limit will prevent the 74188 from overheating and presumably destroy the device.

Switch SW8 is a selector switch that has a one-pole, 10-position format. Only eight locations are needed (B0 to B7) on the switch. So the other two positions can be left unwired. SW8 should have a knob screwed on its shaft that has been numbered to indicate its position. (Knobs of this type are numbered from 0 to 9. SW8 should be wired as to have position 0 connected to B0, position 1 connected to B1, and so on until position 7 is wired to B7.) An example of a substitute knob (if the numbered knob cannot be found), is illustrated in Fig. B-10. Note that a pointing arrow is printed on the *skirt*. If this type of knob is used, the numbers 0 to 7 must be printed on the front panel. For the professional look, use rub-on transfers.Usually, knobs of this type, as well as the numbered transfers, can be found at your local Radio Shack Store or a major electronic mail order house. If a numbered knob is used, attach it to the shaft of SW8 such that the number 0 corresponds to the position B0 on SW8. The other numbers will automatically fall into place if the rotary switch was wired correctly to start with.

Fig. B-10   *By using a skirted knob on the shaft of SW8, you can very easily determine the position of the selector switch.*

Resistors R3 – R7 are 330 Ω ¹/₄ W components. Their main function is to maintain a logic 0 to all address pins of the IC when their associated switch is in the open position.

In Fig. B-11 is a drawing illustrating the recommended housing layout for the completed PROM Programmer. Take note of the special 16-pin IC socket (on top of the housing). Because this programmer will be used to program a large number of ICs, purchase ZIF force, zero insertion, socket from Jameco Electronics or any other fine mail order house (see Fig. B-12). The price of this socket is about $6.00. If you compare this to the price of a standard 16-pin socket (usually $.25), you might be discouraged from using it. But take into account that inexpensive sockets were made to have the IC inserted only once or twice and then left to stay. If you plan to make extensive use of the programmer, the inexpensive sockets will break and need to be replaced frequently. To

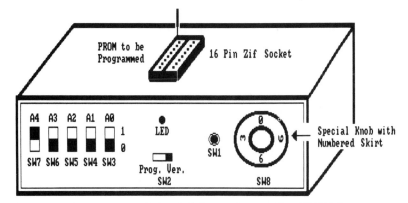

Fig. B-11   *The recommended switch layout for your first PROM Programmer.*

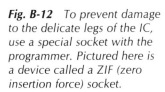

Fig. B-12   *To prevent damage to the delicate legs of the IC, use a special socket with the programmer. Pictured here is a device called a ZIF (zero insertion force) socket.*

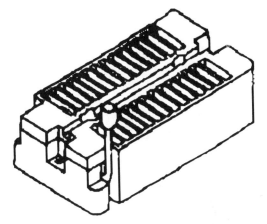

save time and expense, dig down into your pocket and buy the ZIF socket. ZIF sockets allow you to drop the IC to be programmed right into place with no force at all. Then to secure the 16 legs, push a small lever down. When in place, this lever squeezes all the IC legs between the socket contacts. Once programmed, pull the lever upright. This reopens the socket contacts, and the now programmed IC can be removed without the damage you might expect from a standard socket.

EXPERIMENTAL PROGRAMMING SEQUENCE

To program the 74188, just select the address using the SW3 to SW7 switches (A0 – A4). Adjust SW8 to the desired output where a logic 1 is to be placed. You can choose from the following outputs:

B0   B1   B2   B3   B4   B5   B6   B7

Once an output has been selected using SW8, flip SW2 to its program position. Then press SW1. Do not hold SW1 down for more than one second. If you hold it for more than the recommended time limit, you run the risk of damaging the delicate internal electronics of the integrated circuit.

To verify that a logic 1 has been programmed, just flip SW2 to its verify position. When flipped, the LED that is connected to the 1000 Ω resistor will light. If the LED does not light, just repeat the programming process again. Then flip SW2 back into its verify position. This time, the LED should light. If so, you have just programmed a logic 1 to appear at the selected output line every time the associated address is selected, either manually or electronically.

PROGRAMMING THE 74188 TO OPERATE THE DIGITALKER

Now that you know how to program an integrated circuit, proceed to program another IC to provide the binary codes needed to make your Digitalker say a few words. It is easier than you think.

With a copy of Table B-1 in front of you, insert a new 74188 or 8223 PROM into the ZIF socket on top of your programmer housing. Then push the ZIF lever down to secure the IC.

Connect the programmer to its power supply (remember that a power supply that is capable of delivering 300 to 350 mA must be used if you are to be successful).

With power supplied and all switches in a neutral position (neutral position: SW2 to OFF, SW3 – SW7 to logic 0, and SW8 to B0) it is time to program your first IC. Take your time. Verify all switch settings before pressing SW1. Then begin:

1. Address 0 requires a binary output of all 0s, so no programming is needed here.
2. Address 1—flip A4, A3, A2, and A1 to 0.
3.   Flip A0 to 1.
4. Turn SW8 to B0.
5. Flip SW2 to its program position.
6. Press SW1 not for more than one second.
7. Flip SW2 to verify—LED will light. If not, repeat steps 2 to 7.
8. Address 2—Flip A4, A3, A2, and A0 to 0.
9.   Flip A1 to 1.
10. Turn SW8 to B1.
11. Flip SW2 to program, then press SW1.
12. Verify program by flipping SW2 to verify—LED will light.
13. Address 3—Flip A4, A3, and A2 to 0.
14.   Flip A1 and A0 to 1.
15. Flip SW8 to B0.
16. Flip SW2 to program—press SW1—verify logic level.
17. Flip SW8 to B1.
18. Flip SW2 to program—press SW1—verify logic level.
19. Address 4—Flip A4, A3, A1, and A0 to 0.
20.   Flip A2 to 1.
21. Flip SW8 to B2.

22. Flip SW2 to program—press SW1—verify logic level.
23. Address 5—Flip A4, A3, and A1 to 0.
24.   Flip A2 and A0 to 1.
25. Flip SW8 to B0.
26. Flip SW2 to program—press SW1—verify logic level.
27. Flip SW8 to B2.
28. Flip SW2 to program—press SW1—verify logic level.
29. Address 6—Flip A4, A3, and A0, to 0.
30.   Flip A1 and A2 to 1.
31. Flip SW8 to B1.
32. Flip SW2 to program—press SW1—verify logic level.
33. Flip SW8 to B2.
34. Flip SW2 to program—press SW1—verify logic level.
35. Address 7—Flip A4, A3 to 0.
36.   Flip A2, A1, A0 to 1.
37. Flip SW8 to B0.
38. Flip SW2 to program—press SW1—verify logic level.
39. Flip SW8 to B1.
40. Flip SW2 to program—press SW1—verify logic level.
41. Flip SW8 to B2.
42. Flip SW2 to program—press SW1—verify logic level.
43. Address 8—Flip A4, A2, A1, and A0 to 0.
44.   Flip A3 to 1.
45. Flip SW8 to B3.
46. Flip SW2 to program—press SW1—verify logic level.
47. Address 9—Flip A4, A2, A1, and A0 to 0.
48.   Flip A3 and A0 to 1.
49. Flip SW8 to B0.
50. Flip SW2 to program—press SW1—verify logic level.
51. Flip SW8 to B3.
52. Flip SW2 to program—press SW1—verify logic level.
53. Address operator—Flip A4, A2, and A0 to 0.
54.   Flip A1 and A3 to 1.
55. Flip SW8 to B0.
56. Flip SW2 to program—press SW1—verify logic level.
57. Flip SW8 to B1.
58. Flip SW2 to program—press SW1—verify logic level.
59. Flip SW8 to B2.
60. Flip SW2 to program—press SW1—verify logic level.
61. Flip SW8 to B3.
62. Flip SW2 to program—press SW1—verify logic level.
63. Flip SW8 to B4.
64. Flip SW2 to program—press SW1—verify logic level.
65. Flip SW8 to B5.
66. Flip SW2 to program—press SW1—verify logic level.
67. Flip SW8 to B6.

68. Flip SW2 to program—press SW1—verify logic level.
79. Flip SW8 to B7.
70. Flip SW2 to program—press SW1—verify logic level.

This completes the programming of your first PROM.

PARTS LIST FOR PROGRAMMER IN FIG. B-9

R1 R3 R4 R5 R6 R7	330 Ω 1/4 W resistors
R2	1000 Ω 1/4 W resistor
C1	0.47 μF 200V Capacitor
C2 C3	100 μF 35V Electrolytic Capacitor
LED1	Standard red or green LED
IC1	7805 voltage regulator
SW1	SPST push-button switch
SW2	DPDT center-off toggle switch
SW3 – SW7	SPST toggle switches
SW8	1-pole 10-position rotary switch
1	Numbered knob for SW8
1	16-pin ZIF socket (Jameco Electronics)
1	Housing
1	12 Vdc 300 – 350 mA power supply

PROGRAMMER MODIFICATIONS

From the previous step-by-step instructions, you can see that the positioning of the address switches (SW3 – SW7) is an extremely important part of the programming procedure. Just by forgetting to flip, say SW3 to the logic 0 position, you can inadvertently place needed binary material into a location that either does not exist or in the wrong point in the memory. To make the programming of the IC as easy and foolproof as possible, the programmer has a slight modification. The results of these efforts can be seen in Fig. B-13.

The internal electronic wiring is basically the same as shown in Fig. B-9, but there is an additional IC socket (S01). Also, switches SW3 to SW7 have been replaced with two BCD (binary coded decimal) thumbwheel switches. These switches are specially designed devices that provide, at its four output pins (or connector terminals 1, 2, 4, and 8), the binary equivalent of the decimal number that can be seen on the thumbwheel face (see Fig. B-14). For example, say that the number 3 can be seen on the face of the thumbwheel switch. At its output, the switch will provide the binary code of this number (0011) or any other decimal number within the range of 0 to 9 (0, 1, 2, 3, . . . 9). Using

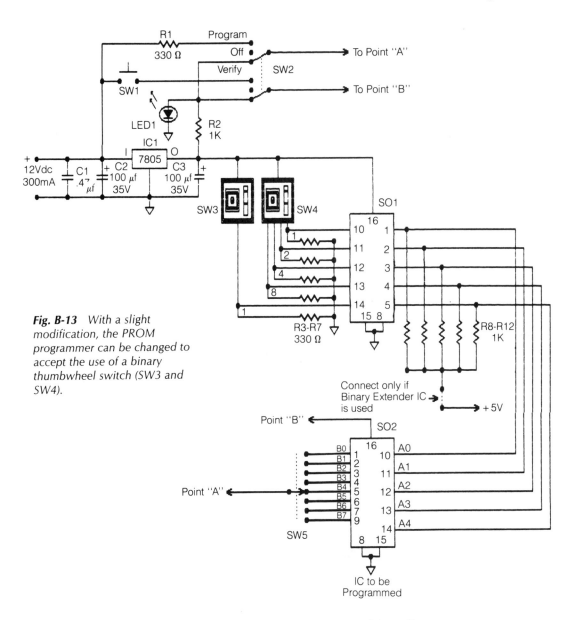

Fig. B-13 With a slight modification, the PROM programmer can be changed to accept the use of a binary thumbwheel switch (SW3 and SW4).

Fig. B-14 A typical thumbwheel switch. Make sure you purchase a switch that has a BCD (binary coded decimal) output.

thumbwheel switches can be the easiest and a foolproof way of entering binary address codes.

Using thumbwheel switches for address input can eliminate complicated switch arrangements, but it does have a drawback. BCD thumbwheel switches are made to provide the binary equivalent of ten numbers only. At the eleventh position, the internal switch arrangement will reset back to the code 0000 instead of advancing to the next standard binary output of 1011 (binary equivalent of the number 11). If there were no way to rectify this problem, you would be able to address only the first 10 addresses instead of the available 32 possible locations.

To overcome this problem, you can, using your PROM programmer, instruct another 74188 IC to make the necessary binary conversions automatically. You can say that you are extending the normal 10-position thumbwheel switch to 20 positions.

CONSTRUCTING THE NEW PROM PROGRAMMER

Figure B-13 illustrates the wiring changes made to the original programmer. In this illustration, you can see that a second 16-pin IC socket is added along with eight 1/4 W carbon resistors. The 330 Ω resistors associated with thumbwheel switches SW3 and SW4 provide a constant ground (or logic 0) connection to S01 pins 10, 11, 12, 13, and 14. Resistors R8 to R12 are 1000 Ω devices that provide the needed voltage drop for the open-collector memory IC (74188/8223). The + 5 V that is brought to the 1000 Ω resistors is not connected at this time, but make the wiring easily accessible for future soldering. As for the schematic in Fig. B-9, S02 is still the recommended 16-pin ZIF socket, which is mounted on top of the selected housing. The new housing layout is shown in Fig. B-17.

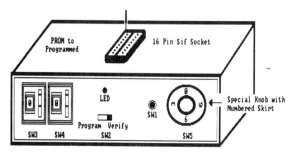

Fig. B-15 *The recommended switch layout for the second PROM programmer.*

If you have never worked with a thumbwheel switch before, Fig. B-15 shows you what you are up against. The back side of both switches are made of a printed circuit board. The desired binary code is etched into the copper of each PC. Connection points for the switches are in the form of PCB connector

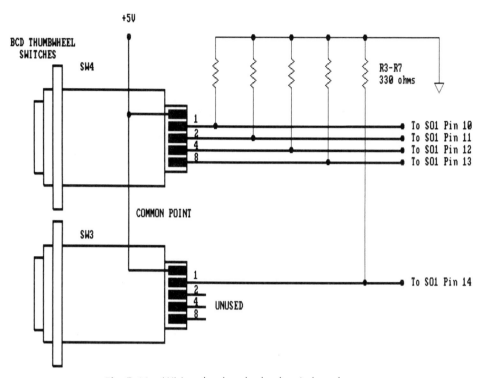

Fig. B-16 *Wiring the thumbwheel switch to the programmer.*

legs, which provide an easy way to mate the switch to the main breadboard either by soldering or purchasing a mating plug. Both switches have the numbering 1, 2, 4, 8, and the word *COMMON* etched into the copper. The numbers 1, 2, 4, and 8 are the designations of the output binary code. The 1 is the D0 output, 2 is the D1, 4 is the D3, and 8 is the D4 output. These four logic outputs should be soldered to their appropriate S01 pins, along with the required 330 Ω resistors. The connection named COMMON is just that—a point common to the output pins. From the schematic, you can see that the common connection legs of both thumbwheel switches are wired together and then soldered to the output of IC1 (+ 5 V voltage regulator). This arrangement provides a correctly switched address data code just by flipping the thumbwheel switch to the desired decimal location.

PROGRAMMING THE BINARY EXTENDER IC

As mentioned, you must program an IC to convert the binary data output of the thumbwheel switches accurately into the standard binary truth table seen in Fig. B-16. But first you must modify your programmer slightly to accommodate this procedure.

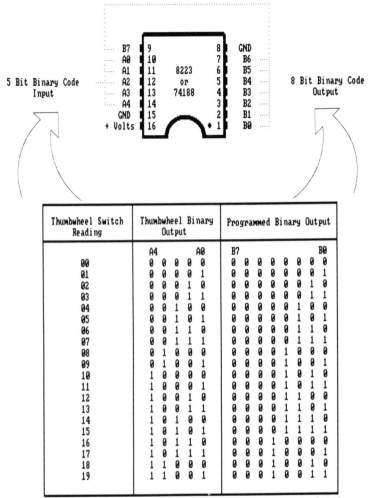

Thumbwheel Switch Reading	Thumbwheel Binary Output					Programmed Binary Output							
	A4				A0	B7							B0
00	0	0	0	0	0	0	0	0	0	0	0	0	0
01	0	0	0	0	1	0	0	0	0	0	0	0	1
02	0	0	0	1	0	0	0	0	0	0	0	1	0
03	0	0	0	1	1	0	0	0	0	0	0	1	1
04	0	0	1	0	0	0	0	0	0	0	1	0	0
05	0	0	1	0	1	0	0	0	0	0	1	0	1
06	0	0	1	1	0	0	0	0	0	0	1	1	0
07	0	0	1	1	1	0	0	0	0	0	1	1	1
08	0	1	0	0	0	0	0	0	0	1	0	0	0
09	0	1	0	0	1	0	0	0	0	1	0	0	1
10	1	0	0	0	0	0	0	0	0	1	0	1	0
11	1	0	0	0	1	0	0	0	0	1	0	1	1
12	1	0	0	1	0	0	0	0	0	1	1	0	0
13	1	0	0	1	1	0	0	0	0	1	1	0	1
14	1	0	1	0	0	0	0	0	0	1	1	1	0
15	1	0	1	0	1	0	0	0	0	1	1	1	1
16	1	0	1	1	0	0	0	0	1	0	0	0	0
17	1	0	1	1	1	0	0	0	1	0	0	0	1
18	1	1	0	0	0	0	0	0	1	0	0	1	0
19	1	1	0	0	1	0	0	0	1	0	0	1	1

NOTE: Each 1 in this column denotes a Blown Fuse in the PROM IC.

Fig. B-17   *To extend the usable BCD output of a thumbwheel switch from its normal 0-to-9 capability to a 00-to-19 capability, you must program a 74188/8223 to compensate for its drawbacks. Above is the truth table for this program.*

PROGRAMMER MODIFICATIONS

Using a length of bare copper wire, made the following connections to the 16-pin IC socket (S01):

1. Short S01 pin 10 to S01 pin 1.
2. Short S01 pin 11 to S01 pin 2.
3. Short S01 pin 12 to S01 pin 3.
4. Short S01 pin 13 to S01 pin 4.
5. Short S01 pin 14 to S01 pin 5.

Double check and make sure that there are no jumper wires touching another jumper. Also check to see that there is no voltage being applied to the five 1000 Ω resistors (R8 to R12).

With the proper power supply connected to the PROM programmer, place a new 74188 or 8223 memory IC into the S02 ZIF socket and secure it into place by lowering the locking lever. When all is connected correctly, refer to the truth table shown in Fig. B-16 and begin your programming.

If you still have doubts in your programming experience, the following step-by-step procedure will assure you of a properly programmed IC. Just take your time and double check all switch positions before pressing SW1.

1. Adjust the thumbwheel switches to read 01.
2. Flip SW8 to the B0 position.
3. Flip SW2 to its program position—press SW1—verify logic level.
4. Adjust the thumbwheel switches to read 02.
5. Flip SW8 to the B1 position.
6. Flip SW2 to its program position—press SW1—verify logic level.
7. Adjust the thumbwheel switches to read 03.
8. Flip SW8 to the B0 position.
9. Flip SW2 to its program position—press SW1—verify logic level.
10. Flip SW8 to the B1 position.
11. Flip SW2 to its program position—press SW1—verify logic level.
12. Adjust the thumbwheel switches to read 04.
13. Flip SW5 to the B2 position.
14. Flip SW2 to its program position—press SW1—verify logic level.
15. Adjust the thumbwheel switches to read 05.
16. Flip SW5 to the B0 position.
17. Flip SW2 to its program position—press SW1—verify logic level.
18. Flip SW5 to the B2 position.
19. Flip SW2 to its program position—press SW1—verify logic level.
20. Adjust the thumbwheel switches to read 06.
21. Flip SW5 to the B1 position.
22. Flip SW2 to its program position—press SW1—verify logic level.
23. Flip SW5 to the B2 position.
24. Flip SW2 to its program position—press SW1—verify logic level.
25. Adjust the thumbwheel switches to read 07.
26. Flip SW5 to the B0 position.
27. Flip SW2 to its program position—press SW1—verify logic level.
28. Flip SW5 to the B1 position.
29. Flip SW2 to its program position—press SW1—verify logic level.
30. Flip SW5 to the B2 position.
31. Flip SW2 to its program position—press SW1—verify logic level.
32. Adjust the thumbwheel switches to read 08.
33. Flip SW5 to the B3 position.
34. Flip SW2 to its program position—press SW1—verify logic level.
35. Adjust the thumbwheel switches to read 09.

36. Flip SW5 to the B0 position.
37. Flip SW2 to its program position—press SW1—verify logic level.
38. Flip SW5 to the B3 position.
39. Flip SW2 to its program position—press SW1—verify logic level.
40. Adjust the thumbwheel switches to read 10.
41. Flip SW5 to the B1 position.
42. Flip SW2 to its program position—press SW1—verify logic level.
43. Flip SW5 to the B3 position.
44. Flip SW2 to its program position—press SW1—verify logic level.
45. Adjust the thumbwheel switches to read 11.
46. Flip SW5 to the B0 position.
47. Flip SW2 to its program position—press SW1—verify logic level.
48. Flip SW5 to the B1 position.
49. Flip SW2 to its program position—press SW1—verify logic level.
50. Flip SW5 to the B3 position.
51. Flip SW2 to its program position—press SW1—verify logic level.
52. Adjust the thumbwheel switches to read 12.
53. Flip SW5 to the B2 position.
54. Flip SW2 to its program position—press SW1—verify logic level.
55. Flip SW5 to the B3 position.
56. Flip SW2 to its program position—press SW1—verify logic level.
57. Adjust the thumbwheel switches to read 13.
58. Flip SW5 to the B0 position.
59. Flip SW2 to its program position—press SW1—verify logic level.
60. Flip SW5 to the B2 position.
61. Flip SW2 to its program position—press SW1—verify logic level.
62. Flip SW5 to the B3 position.
63. Flip SW2 to its program position—press SW1—verify logic level.
64. Adjust the thumbwheel switches to read 14.
65. Flip SW5 to the B1 position.
66. Flip SW2 to its program position—press SW1—verify logic level.
67. Flip SW5 to the B2 position.
68. Flip SW2 to its program position—press SW1—verify logic level.
69. Flip SW5 to the B3 position.
70. Flip SW2 to its program position—press SW1—verify logic level.
71. Adjust the thumbwheel switches to read 15.
72. Flip SW5 to the B0 position.
73. Flip SW2 to its program position—press SW1—verify logic level.
74. Flip SW5 to the B1 position.
75. Flip SW2 to its program position—press SW1—verify logic level.
76. Flip SW5 to the B2 position.
77. Flip SW2 to its program position—press SW1—verify logic level.
78. Flip SW5 to the B3 position.
79. Flip SW2 to its program position—press SW1—verify logic level.
80. Adjust the thumbwheel switches to read 16.
81. Flip SW5 to the B4 position.

82. Flip SW2 to its program position—press SW1—verify logic level.
83. Adjust the thumbwheel switches to read 17.
84. Flip SW5 to the B0 position.
85. Flip SW2 to its program position—press SW1—verify logic level.
86. Flip SW5 to the B4 position.
87. Flip SW2 to its program position—press SW1—verify logic level.
88. Adjust the thumbwheel switches to read 18.
89. Flip SW5 to the B1 position.
90. Flip SW2 to its program position—press SW1—verify logic level.
91. Flip SW5 to the B4 position.
92. Flip SW2 to its program position—press SW1—verify logic level.
93. Adjust the thumbwheel switches to read 19.
94. Flip SW5 to the B0 position.
95. Flip SW2 to its program position—press SW1—verify logic level.
96. Flip SW5 to the B1 position.
97. Flip SW2 to its program position—press SW1—verify logic level.
98. Flip SW5 to the B4 position.
99. Flip SW2 to its program position—press SW1—verify logic level.

Because advancing the thumbwheel switches to the 20 position would require a memory IC making use of the six-bit data input. This is, of course, impossible with the available 74188, so you must stop at this point.

If your design requires you to address data above the limit set by the thumbwheel switches, you might want to reconsider rewiring the programmer and stick with the original design. If this is your decision, pay strict attention to the setting of the address switches. You do not want to ruin the IC.

INSTALLATION OF THE PROGRAMMED BINARY EXTENDER IC

With all 20 positions programmed into the PROM IC, it is time to install it inside your programmer. To prevent any accidental electrical shock or damage to your IC, remove the programmer from the power supply. When power is removed, the previously installed S01 jumpers can also be removed. Now make the connection of the +5 V to resistors R8 to R12. For help in locating this point, refer back to Fig. B-13.

Now install the newly programmed PROM into the S01 socket, being careful to orient the IC properly before seating the IC in the socket.

With the IC installed, you have a sophisticated piece of electronic gear.

ADDITIONAL MODIFICATION TO THE PROM PROGRAMMER

For all those of you that want the ultimate in electronic gear, Fig. B-18 presents another modification that can be made to PROM programmer, and Fig. B-19 presents a suggested housing layout.

Figure B-18 shows that the normal one-pole, 10-position rotary switch used for SW8 in Fig. B-8 (SW5 in Fig. B-13) can be replaced with the more pleasing thumbwheel switch. Unlike SW3 and SW4 (BCD thumbwheel

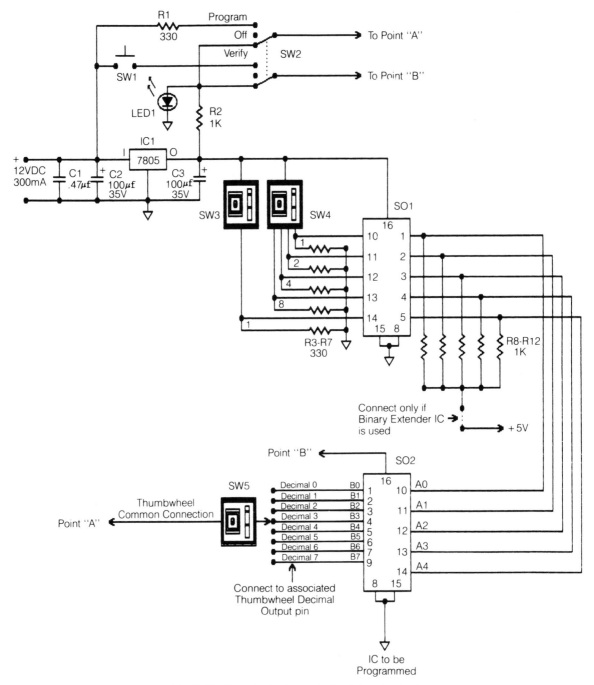

Fig. B-18 *By substituting a decimal thumbwheel*
for the rotary switch (SW5), you have the makings of the third PROM programmer.

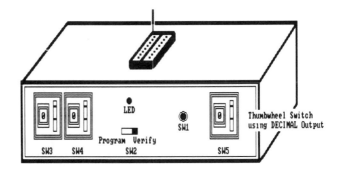

SW3 and SW4 - Thumbwheel Switches using
BCD Output

Fig. B-19   *The recommended switch layout for the third PROM programmer.*

switches), SW5 provides the same standard connections that were found with the SW5 rotary switch. The difference is the physical appearance.

If you wish to purchase this new addition to the programmer, Jameco, Digi-Key, and All Electronics have both the BCD (SW3 and SW4 thumbwheel switches) and the one-pole, 10-position decimal thumbwheel switch at reasonable prices. Why not take the plunge and build the ultimate in PROM programmers? It will take only a small cash investment on your part.

PARTS LIST FOR THE PROGRAMMER IN FIG. B-13

R1 R3 R4 R5 R6 R7	330 Ω $1/4$ W resistors
R2 R8 R9 R10 R11 R12	1000 Ω $1/4$ W resistors
C1	0.47 μF 200 V capacitor
C2 C3	100 μF 35 V electrolytic capacitor
LED1	Standard red or green LED
IC1	7805 voltage regulator
SW1	SPST push-button switch
SW2	DPDT center-off toggle switch
SW3 SW4	BCD thumbwheel switches
SW5	One-pole 10-position rotary switch
1	Numbered knob for SW5
S02	16-pin ZIF socket (Jameco Electronics)
S01	Standard 16-pin IC socket
1	Housing
1	12 Vdc 300 – 350 mA power supply

PARTS LIST FOR THE PROGRAMMER IN FIG. B-18

R1 R3 R4 R5 R6 R7	330 Ω 1/4 W resistors
R2 R8 R9 R10 R11 R12	1000 Ω 1/4 W resistors
C1	0.47 μF 200 V capacitor
C2 C3	100 μF 35 V electrolytic capacitor
LED1	Standard red or green LED
IC1	7805 voltage regulator
SW1	SPST push-button switch
SW2	DPDT center-off toggle switch
SW3 SW4	BCD thumbwheel switches
SW5	Decimal thumbwheel switch—one-pole, 10-position rotary switch
S02	16-pin ZIF socket (Jameco Electronics)
S01	Standard 16-pin IC socket
1	Housing
1	12 Vdc 300 – 350 mA power supply

☏C
The Schematics, the Parts, and the Phones

THIS APPENDIX PROVIDES EXPLODED VIEWS, SCHEMATICS, AND PARTS LISTS FOR MOST commonly used home telephones.

THE 30G ROTARY DIAL

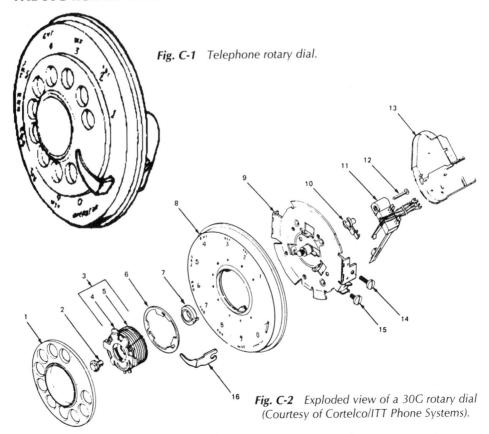

Fig. C-1   *Telephone rotary dial.*

Fig. C-2   *Exploded view of a 30G rotary dial (Courtesy of Cortelco/ITT Phone Systems).*

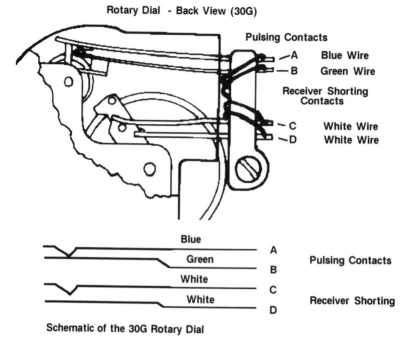

Rotary Dial - Back View (30G)

Pulsing Contacts

A   Blue Wire
B   Green Wire

Receiver Shorting Contacts

C   White Wire
D   White Wire

Blue
Green

White
White

A
B
C
D

Pulsing Contacts

Receiver Shorting

Schematic of the 30G Rotary Dial

Fig. C-3 *Back view of a rotary dial showing pulsing and receiver-shorting contacts.*

Table C-1. Parts list for the 30G rotary dial.

Parts List for the 30G Rotary Dial

INDEX NUMBER	DESCRIPTION
1	Plate, Finger
2	Screw, Hex-Head
3	Spring and Spider Assembly
4	Spider
5	Spring
6	Ring, Retaining
7	Bushing
8	Ring, Numeral
9	Gear Train and Bracket Assembly
10	Actuator
11	Contact Spring Assembly
12	Screw, Contact Spring Mounting
13	Cover, Dust
14	Screw, Finger Stop Mounting
15	Screw, Dial Mounting
16	Finger Stop

THE 42 opg TONE DIAL

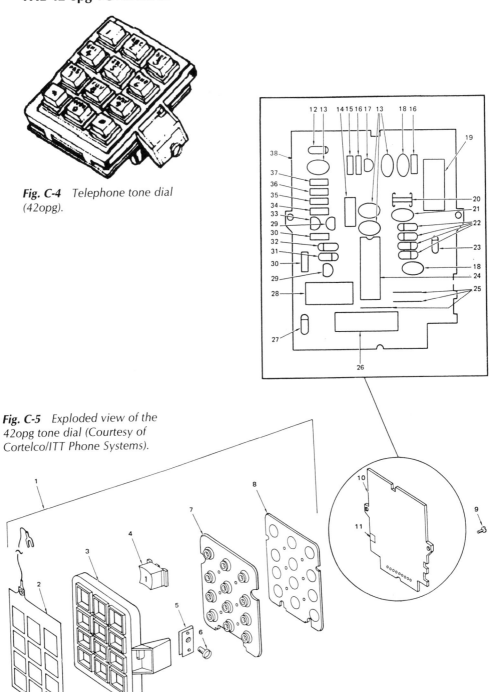

Fig. C-4 *Telephone tone dial (42opg).*

Fig. C-5 *Exploded view of the 42opg tone dial (Courtesy of Cortelco/ITT Phone Systems).*

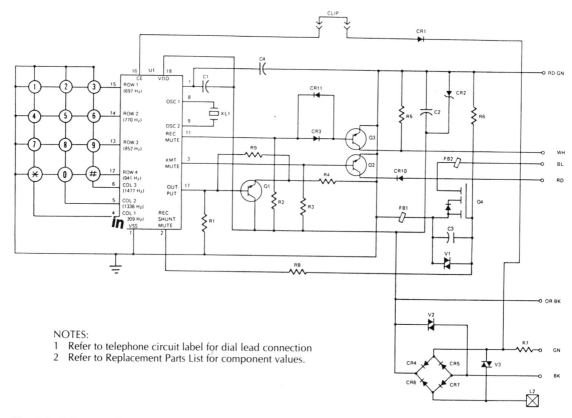

Fig. C-6   *Schematic diagram of the 42opg tone dial (Courtesy of Cortelco/ITT Phone Systems).*

NOTES:
1   Refer to telephone circuit label for dial lead connection
2   Refer to Replacement Parts List for component values.

Table C-2. Parts list for the 42opg tone dial.

Parts List for the 42 opg Tone Dial	
INDEX NUMBER	DESCRIPTION
1	Keypad Assembly
2	Shield, Electrostatic
3	Plate, Cover
4	Plastic Pushbuttons
5	"U"-Nut
6	Screw, Dial Mounting
7	Switchplate, Silicone
8	Printed Circuit Board Assembly #1
9	Screw, PCB Mounting
10	Printed Circuit Board Assembly #2
11	Clip
12	Diode, Zener, 1N4742, CR2
13	Capacitor, .0068uf, C1-C4

INDEX NUMBER	DESCRIPTION
14	Resistor, 2K, R8
15	Resistor, 470K, R6
16	Bead, Ferrite, FB1, FB2
17	Transistor, BS170 ,Q4
18	Varistor, ERZ-C10-DK-180, V1, V2
19	Resistor 10 Ohms, 5W, R7
20	Terminal, Spade
21	Varistor, ERZ-C14-DK-180, V3
22	Diode, SD103, CR4-CR7
23	Diode, 1N4004, CR
24	IC, DTMF Tone Generator, U1
25	Strap, Wire
26	Connector
27	Diode, 1N4448, CR10
28	Crystal, 3.58MHz., XL1
29	Transistor, MPS8092, Q2, Q3
30	Resistor, 3.3K, R2, R3
31	Diode, SD164D, CR11
32	Diode, 1N4148, CR3
33	Transistor, 2N4141, Q1
34	Resistor, 470 Ohms, R9
35	Resistor, 5.1K, R5
36	Resistor, 10K, R1
37	Resistor, 100 Ohms, R4
38	PC Board, Drilled

THE 32G TONE DIAL

Fig. C-7 *Telephone tone dial (32g).*

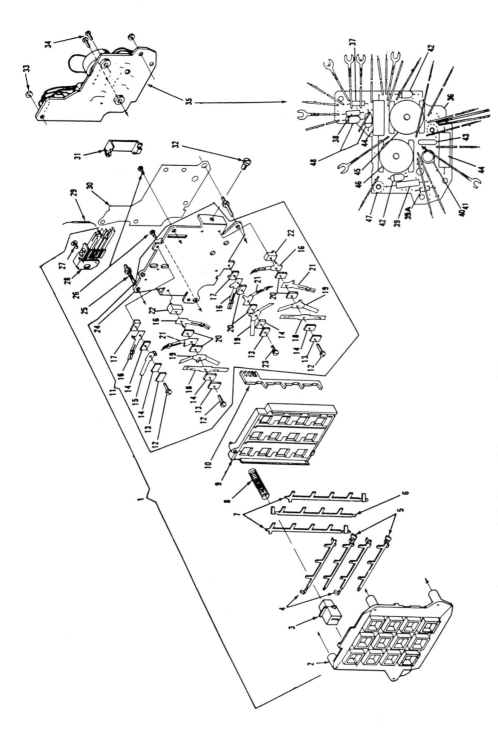

Fig. C-8 Exploded view of the 32g tone dial (Courtesy of Cortelco/ITT Phone Systems).

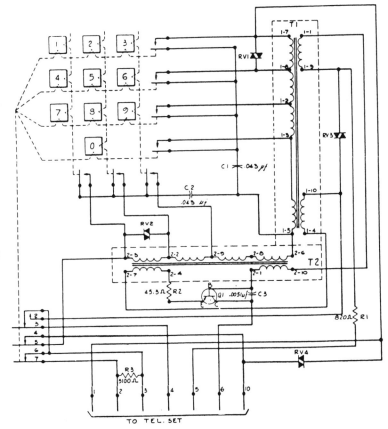

Dotted lines represent Cranks and Slide. Springs 6-7 break first, then in succession the 1-2 springs make, then the 1-3 springs break. The 4-5 springs break last. (The 8 spring, not shown, is a stiffening spring.)

Fig. C-9   *Schematic diagram of the 32g tone dial (Courtesy of Cortelco/ITT Phone Systems).*

Table C-3. Parts list for the 32g tone dial.

Parts List for the 32g Tone Dial	
INDEX NUMBER	DESCRIPTION
1	Pushbutton Assembly
2	Plate
3	Pushbuttons (Metropolitan)
4	Crank, Horizontal (#1 and #3)
5	Crank, Horizontal (#2 and #4)
6	Crank, Vertical (Center)
7	Crank, Vertical (Side)
8	Spring, Push Button
9	Frame, Push Button
10	Slide
11	Mounting Plate Assembly
12	Screw
13	Plate, Clamp
14	Insulator #1
15	Spring

INDEX NUMBER	DESCRIPTION
16	Spring
17	Spacer
18	Insulator
19	Spring
20	Insulator #2
21	Spring
22	Spacer
23	Screw
24	Plate, Mounting
25	Screw, Special
26	Screw, Self-Tapping
27	Screw
28	Spring Assembly
29	Wire (White/Blue/Green)
30	Insulator #3
31	Cover, Spring Assembly
32	Screw, Dial Mounting
33	Locknut
34	Screw, Circuit Board to Mounting Plate
35	Circuit Board Assembly
36	Board, Printed Circuit
37	Resistor, 5.1K, R3
38	Resistor, 820 Ohms, R1
39	Resistor, 45.3 Ohm, R2
40	Transistor
41	Spacer, Transistor
42	Diode, D1, D2
43	Capacitor #1
44	Capacitor #2
45	Transformer (T1)
46	Transformer (T2)
47	Spacer
48	Diode, D3, D4

THE TELEPHONE HANDSET

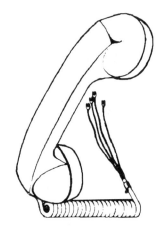

Fig. C-10   *Standard telephone handset.*

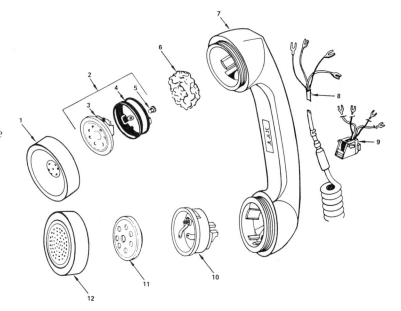

Fig. C-11 *Exploded view of the telephone handset (Courtesy of Cortelco/ITT Phone Systems).*

Table C-4. Parts list for the telephone handset.

Parts List for a Telephone Handset	
INDEX NUMBER	DESCRIPTION
1	Cap, Receiver
2	Receiver and Induction Coil Assembly
3	Receiver
4	Coil Induction
5	Screw, Mounting
6	Baffle, Cotton Ball
7	Handle
8	Cord, Coiled (Spade Lug)
9	Modular Jack
10	Transmitter Holder Assembly
11	Transmitter
12	Cap, Transmitter

THE NETWORK

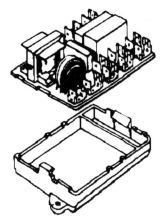

Fig. C-12   *Telephone network (printed circuit board model).*

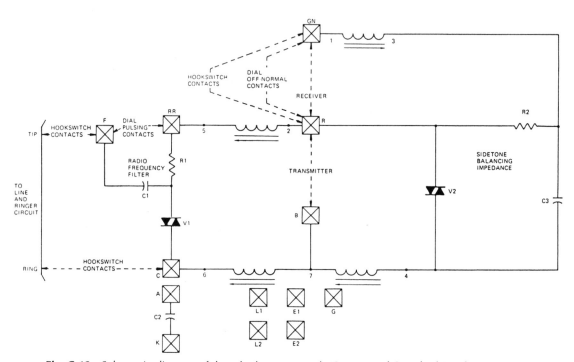

Fig. C-13   *Schematic diagram of the telephone network (Courtesy of Cortelco/ITT Phone Systems).*

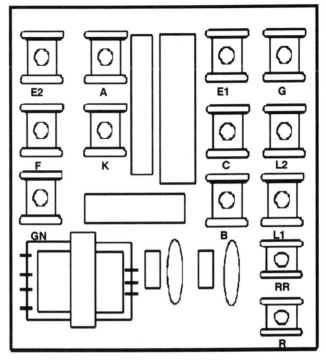

Fig. C-14   *Telephone network terminal designations.*

Fig. C-15   *Telephone network (box type).*

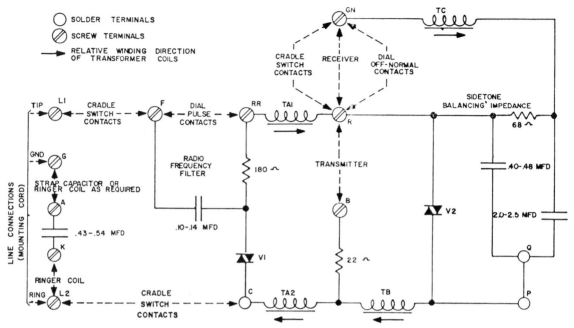

Fig. C-16   Schematic diagram of the older box-type telephone network (Courtesy of Cortelco/ITT Phone Systems).

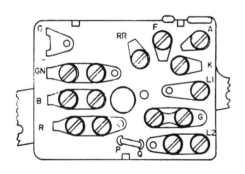

Fig. C-17   Telephone network terminal designations (Courtesy of Cortelco/ITT Phone Systems).

THE 130 BA RINGER

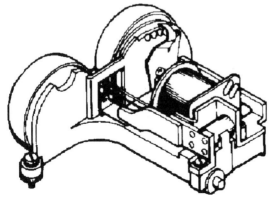

Fig. C-18   Telephone ringer (130BA).

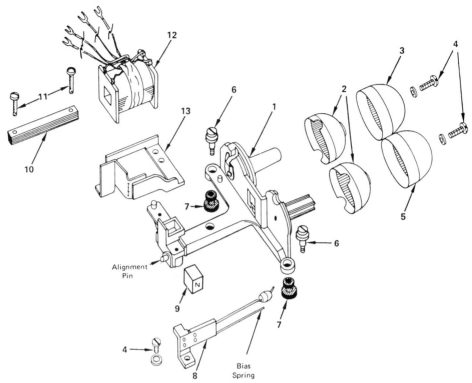

Fig. C-19 *Exploded view of the 130BA ringer (Courtesy of Cortelco/ITT Phone Systems).*

Table C-5. Parts list for the 130BA ringer.

Parts List for the 130 BA Ringer	
INDEX NUMBER	DESCRIPTION
1	Frame
2	Resonator
3	Gong "A"
4	Screw, Mounting
5	Gong "B"
6	Screw, Frame Mounting
7	Grommet, Rubber
8	Armature and Clapper Assembly
9	Magnet
10	Laminations, Core
11	Screw, Laminations Mounting
12	Coil Assembly
13	Support Pole Piece Assembly

THE 148 BA RINGER

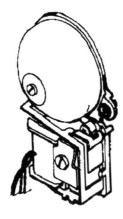

Fig. C-20 *Telephone ringer (148BA).*

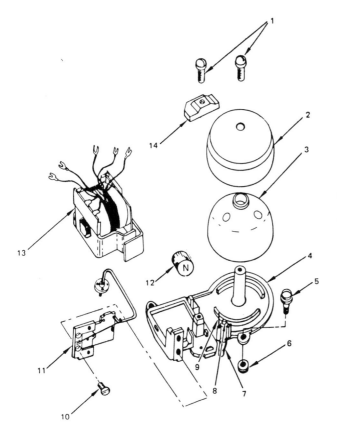

Fig. C-21 *Exploded view of the 148BA ringer (Courtesy of Cortelco/ITT Phone Systems).*

Table C-6. Parts list for the 148BA ringer.

Parts List for the 148 BA Ringer	
INDEX NUMBER	DESCRIPTION
1	Lockwasher Screw
2	Gong
3	Resonator
4	Frame, Ringer Mounting
5	Screw (Ringer Mounting)
6	Grommet
7	Lever, Tone
8	Nut, Push
9	Rod, Stop
10	Screw (Armature and Clapper Assembly Mounting)
11	Armature and Clapper Assembly
12	Magnet
13	Pole Pieces Assembly
14	Plate, Retainer

THE 500 TELEPHONE

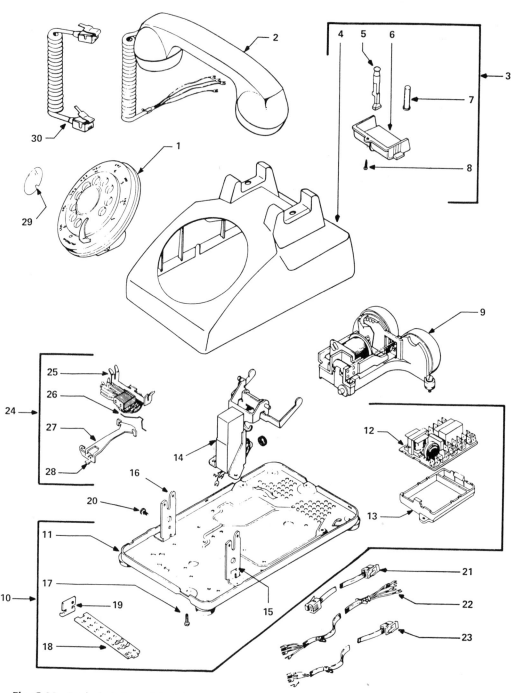

Fig. C-22 *Exploded view of the 500 rotary telephone (Courtesy of Cortelco/ITT Phone Systems).*

K-500 TYPE TELEPHONE CIRCUIT

183078-101

ISSUE NO. 2

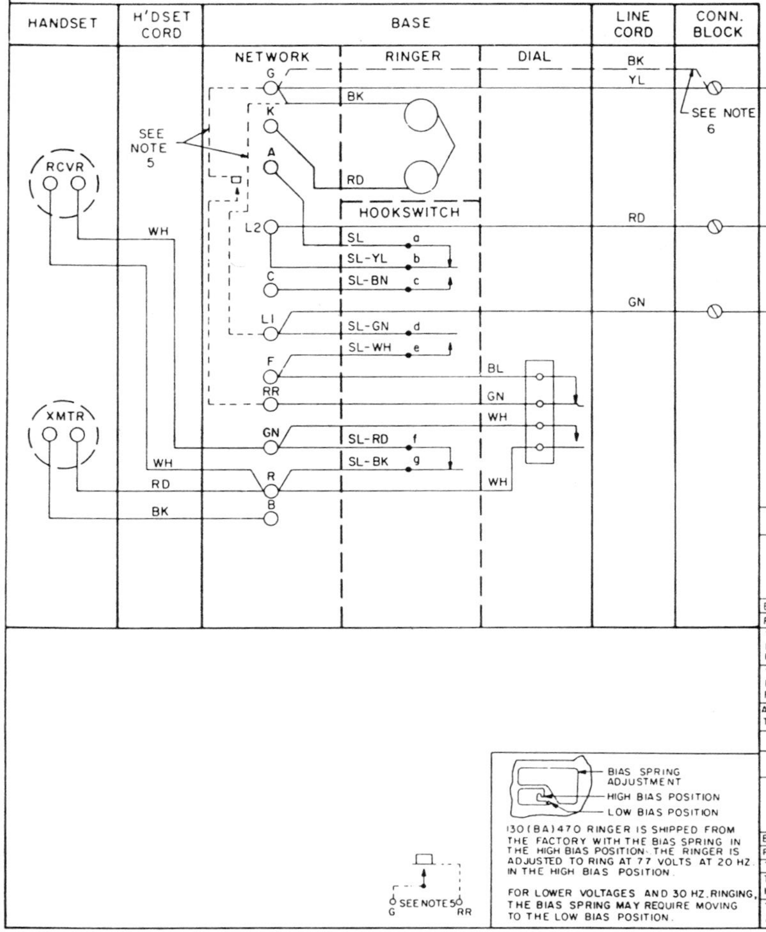

NOTES

1. FOR MANUAL SERVICE: REPLACE DIAL WITH DUMMY PLUG ASSEMBLY AND TRANSFER SL-WH SWITCH LEAD TO "RR" ON NETWORK.

2. TO PERMANENTLY SILENCE RINGER: FOR BRIDGED, RING PARTY AND TIP PARTY EXCEPT DIAL MESSAGE RATE SERVICES, TRANSFER BK RINGER LEAD TO "A" TERMINAL ON NETWORK. FOR TIP PARTY DIAL MESSAGE RATE SERVICE, TRANSFER SL-RD RINGER LEAD TO THE "K" TERMINAL BK LEAD TO "G" AND SL LEAD TO "B" MUST REMAIN CONNECTED. FOR PARTY IDENTIFICATION, FOR AUTOMATIC TICKETING, TRANSFER BK RINGER LEAD TO THE "K" TERMINAL. FOR FREQUENCY SELECTIVE RINGERS, TRANSFER RD RINGER LEAD FROM L2 TO "K" ON NETWORK

3. RINGER CUT-OFF CONTROL BY CUSTOMER. BEND STOP NEXT TO DETENT ON RINGER VOLUME CONTROL SO THAT IT COMPLETELY CLEARS THE RIM OF THE RINGER FRAME. THIS PROVIDES A FURTHER POSITION ON VOLUME CONTROL WHICH PREVENTS ARMATURE MOVEMENT.

4. WHEN THE HANDSET IS REMOVED CONTACT *q f* BREAKS LAST.

5. DOTTED LINE INDICATE GND. CIRCUIT FOR TELEPHONES WITH SPECIAL FEATURE CODE 34.

6. WHEN 4 CONDUCTORS ARE FURNISHED WITH MOUNTING CORD, TERMINATE BK CONDUCTOR WITH YL CONDUCTOR AT CONNECTING BLOCK EXCEPT WHERE USED FOR SPECIAL APPLICATIONS.

7. **IF GND. WIRE IS BROUGHT TO CONNECTING BLOCK.**

8. **TRANSFER SL SWITCH LEAD FROM L2 TO "A" TERMINAL ON NETWORK.**

9. **TAPED AND STORED EXCEPT FOR 16-25 HZ RINGERS WHICH CONNECT SAME AS BIASED RINGERS.**

10. **CONNECTIONS FOR BRIDGED AND RING PARTIES ARE FOR FLAT AND MESSAGE RATE SERVICE.**

11. TABLE A APPLIES TO 4 WIRE 130(OBA) ONLY.

TABLE A — CONNECTIONS FOR BIASED RINGERS (SEE NOTES 1 TO 4)

CLASS OF SERVICE	CONNECTIONS AT CONNECTING BLOCK						CONNECTIONS AT NETWORK				
	LINE		MTG. CORD				RINGER LEADS				
	RING	TIP	GND	RD	GN	YL	RD	BK	SL	SL-RD
BRIDGED 10★	R	G	Y★	G	G	Y	L2	G	K	A
RING PARTY 10★	R	G	Y	G	R	Y	L2	G	K	A
TIP PARTY EXCEPT DIAL MESSAGE RATE	R	G	Y	G	R	Y	L2	G	K	A
TIP PARTY DIAL MESSAGE RATE 8★	R	G	Y	G	R	Y	L2	L1	B	B
AUTOMATIC TICKETING 8★	R	G	Y	G	R	Y	L2	L1	K	B

TABLE B — CONNECTIONS FOR 156-157 FREQUENCY SELECTIVE RINGER

CLASS OF SERVICE	CONNECTIONS AT CONNECTING BLOCK						CONNECTIONS IN SET				
	LINE		MTG CORD				RINGER LEADS				
	RING	TIP	GND	RD	GN	YL	RD	BK	SL	SL-RD
BRIDGED	R	G	Y 7★	R	G	Y	L2	G	9★	9★
RING PARTY	R	G	Y	R	R	Y	L2	G	9★	9★
TIP PARTY	R	G	Y	R	R	Y	L2	G	9★	9★
TIP PARTY DIAL MESSAGE RATE 1000 OHM	R	G	Y	G	R	Y	G	A	9★	B
TIP PARTY DIAL MESSAGE RATE 2650 OHMS B ★	R	G	Y	G	R	Y	G	A	9★	B

HANDSET | H'DSET CORD | NETWORK | BASE | RINGER | HOOKSWITCH | DIAL | LINE CORD | CONN. BLOCK

130(BA)47Q RINGER IS SHIPPED FROM THE FACTORY WITH THE BIAS SPRING IN THE HIGH BIAS POSITION. THE RINGER IS ADJUSTED TO RING AT 77 VOLTS AT 20 HZ IN THE HIGH BIAS POSITION.

FOR LOWER VOLTAGES, AND 30 HZ RINGING, THE BIAS SPRING MAY REQUIRE MOVING TO THE LOW BIAS POSITION.

Fig. C-23   Schematic diagram of the 500 telephone (Courtesy of Cortelco/ITT Phone Systems).

Table C-7. Parts list for the 500 rotary telephone.

Parts List for the 500 Telephone	
INDEX NUMBER	DESCRIPTION
1	Dial Assembly
2	Handset Assembly
3	Housing and Plunger Assembly
4	Housing
5	Plunger (Used for Exclusion Option)
6	Plunger, Retainer
7	Plunger, Cradle Switch
8	Screw, Plunger Retainer
9	130-BA Ringer
10	Base Assembly
11	Plate, Base
12	Printed Circuit, Network
13	Spacer, Network
14	Cradle Switch Assembly
15	Bracket, Dial (R.H.)
16	Bracket, Dial (L.H.)
17	Screw, Cabinet Lock
18	Terminal Board Assembly
19	Mounting Plate
20	Screw
21	Cord, Mounting, Full Modular
22	Cord, Mounting, Standard Spade Lug
23	Cord, Mounting, Quarter Modular
24	Exclusion, Switch Assembly (Optional)
25	Exclusion Switch (Optional)
26	Retainer
27	Bracket
28	Screw, B.H.M.
29	Card, Number, (Circular)
30	Cord, Handset, Modular

THE 2500 TELEPHONE

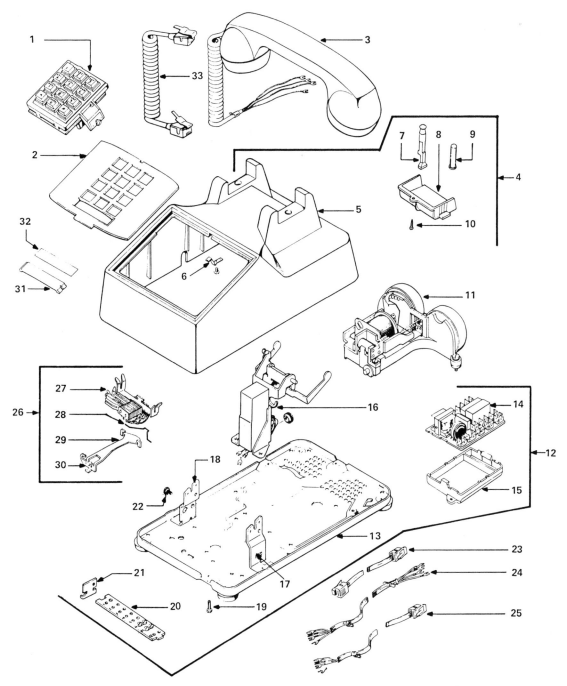

Fig. C-24 *Exploded view of the 2500 telephone (Courtesy of Cortelco/ITT Phone Systems).*

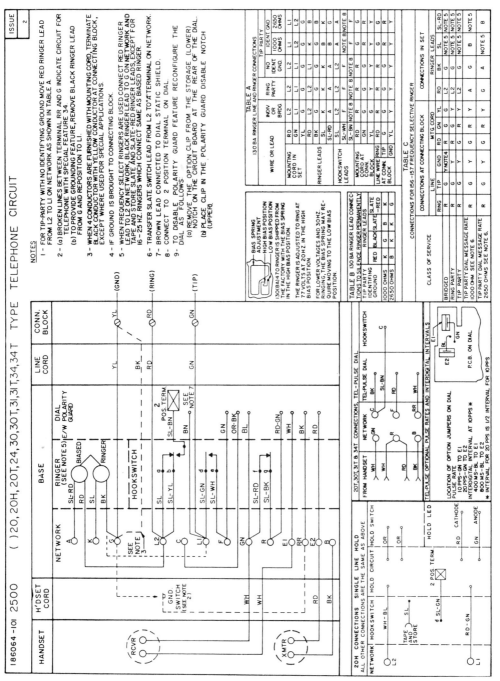

Fig. C-25 Schematic diagram of the 2500 telephone using the newer PC board network (Courtesy of Cortelco/ITT Phone Systems).

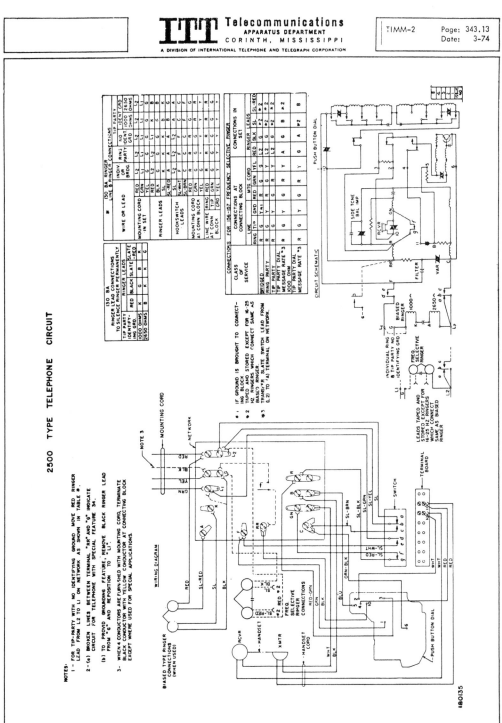

Fig. C-26 Schematic diagram of the 2500 telephone using the older box network (Courtesy of Cortelco/ITT Phone Systems).

Table C-8. Parts list for the 2500 tone-dial telephone.

Parts List for the 2500 Telephone	
INDEX NUMBER	DESCRIPTION
1	42-opg Tone Dial Assembly
2	Face Plate, Pushbutton Dial
3	Handset Assembly
4	Housing and Plunger Assembly
5	Housing
6	Clip, Face Plate
7	Plunger, Exclusion (Optional)
8	Plunger, Retainer
9	Plunger, Cradle Switch
10	Screw, Plunger Retainer
11	130-BA Ringer
12	Base Assembly
13	Plate, Base
14	Printed Circuit, Network
15	Spacer
16	Cradle Switch Assembly
17	Bracket, Dial (R.H.)
18	Bracket, Dial (L.H.)
19	Screw, Cabinet Locking
20	Terminal Board Assembly
21	Mounting Plate
22	Screw
23	Cord Mounting, Full Modular
24	Cord Mounting, Standard Spade Lugs
25	Cord Mounting, Quarter Modular
26	Exclusion Switch Assembly (Optional)
27	Exclusion Switch (Optional)
28	Retainer
29	Bracket
30	Screw
31	Retainer, Number Card
32	Card, Number
33	Cord, Handset Modular

THE 3554 TELEPHONE

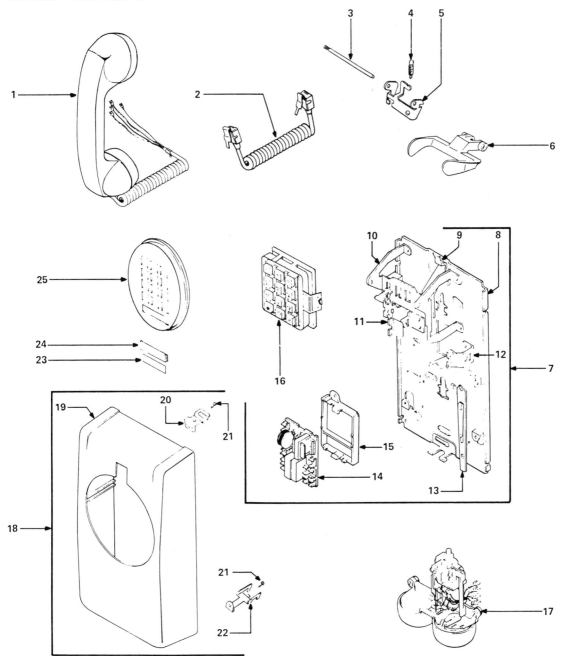

Fig. C-27 *Exploded view of the 3554 telephone (Courtesy of Cortelco/ITT Phone Systems).*

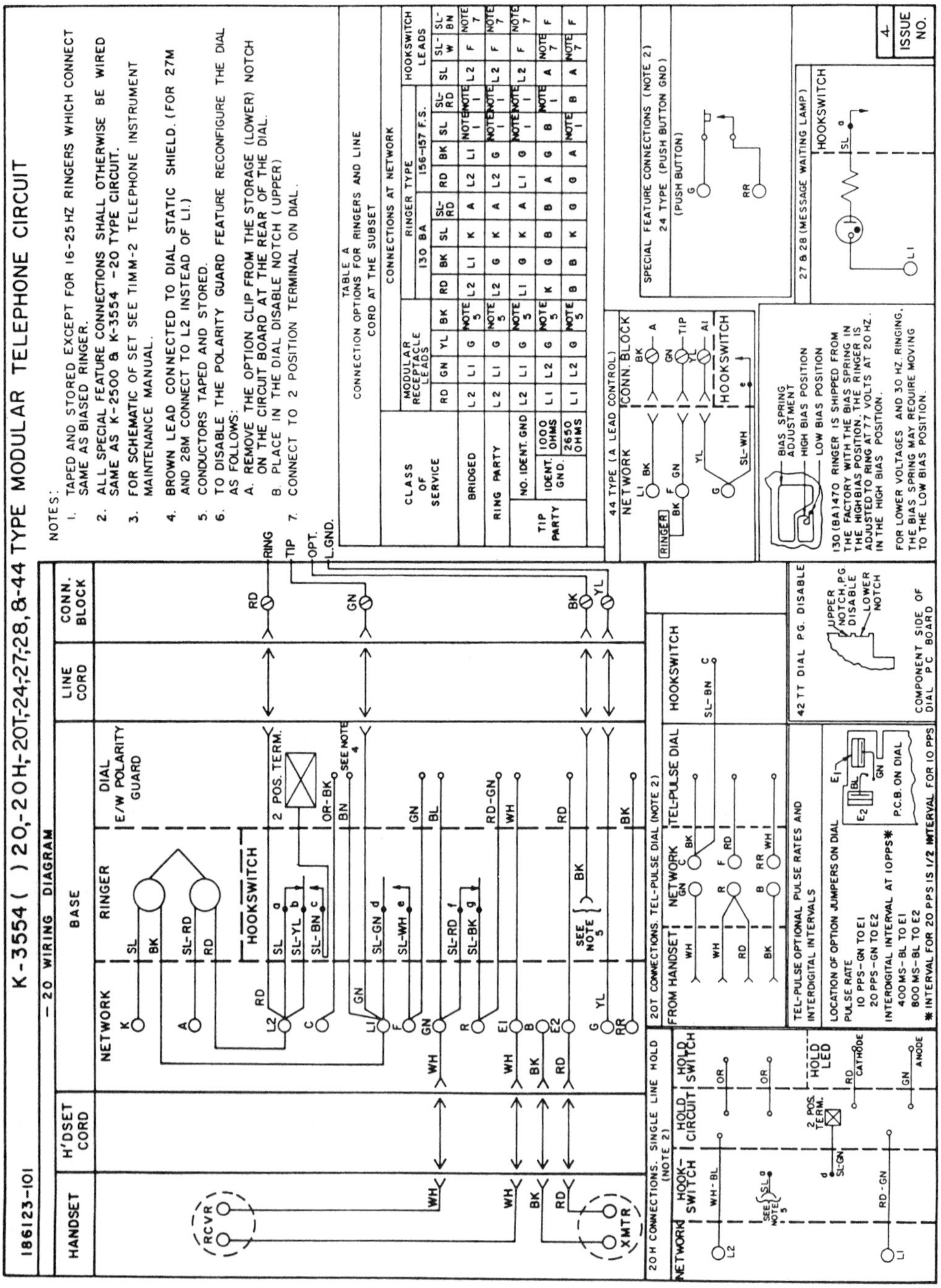

Fig. C-28 Schematic diagram of the 3554 Wall Telephone (Courtesy of Cortelco/ITT Phone Systems).

Table C-9. Parts list for the 3554 telephone.

Parts List for the 3554 Telephone	
INDEX NUMBER	DESCRIPTION
1	Handset
2	Cord, Modular
3	Shaft, Handset Assembly
4	Spring
5	Bracket, Pivot
6	Cradle, Handset
7	Base Assembly
8	Plate, Base
9	Hook, Latch
10	Cradle Switch Assembly
11	Bracket, Dial (L.H.)
12	Bracket, Dial (R.H.)
13	Arm, Ringer Volume Control
14	Printed Circuit Network
15	Spacer
16	42-opg Tone Dial Assembly
17	130-BA Ringer
18	Housing Assembly
19	Housing
20	Bracket, Latch
21	Screw, Latch Bracket and Catch
22	Catch
23	Card, Number
24	Retainer, Number Card
25	Face Plate

THE 2554 TELEPHONE

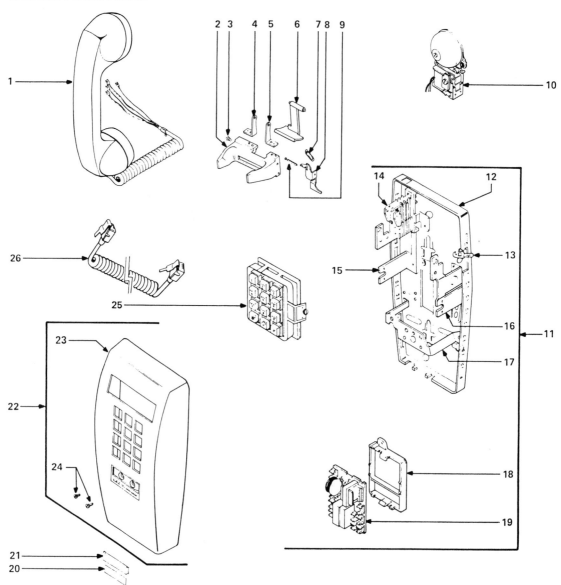

Fig. C-29 *Exploded view of the 2554 telephone (Courtesy of Cortelco/ITT Phone Systems).*

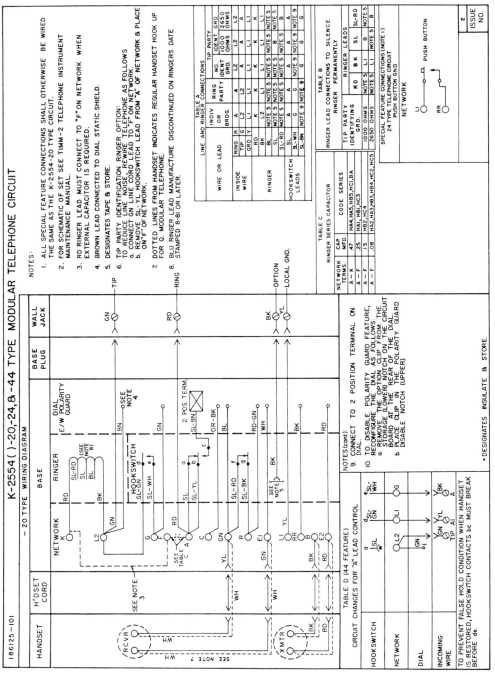

Fig. C-30 *Schematic diagram of the 2554 telephone using the newer PC board network (Courtesy of Cortelco/ITT Phone Systems).*

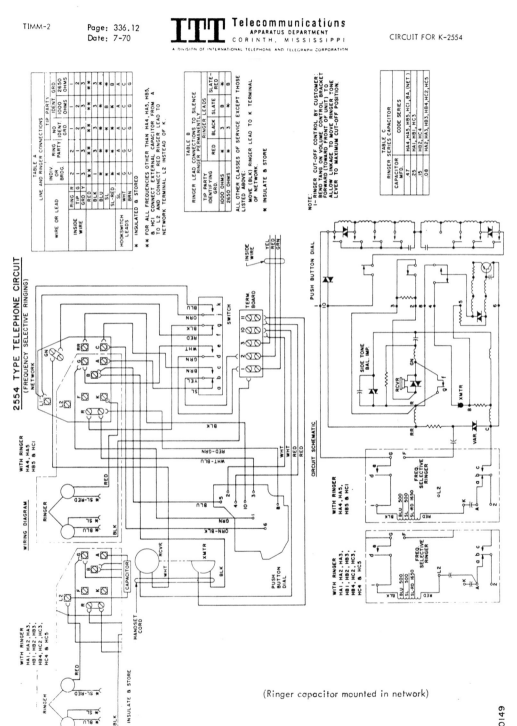

Fig. C-31 Schematic diagram of the 2554 telephone using the older box network (Courtesy of Cortelco/ITT Phone Systems).

Table C-10. Parts list for the 2554 telephone.

Parts List for the 2554 Telephone	
INDEX NUMBER	DESCRIPTION
1	Handset
2	Cradle, Handset
3	Screw, Handset Cradle
4	Bracket, Cradle Plunger (L.H.)
5	Bracket, Cradle Plunger (R.H.)
6	Plunger, Cradle Switch
7	Spring, Cradle Switch
8	Arm, Lever
9	Pin, Pivot
10	148-BA Ringer
11	Base Assembly
12	Plate, Base
13	Level, Ringer Volume Control
14	Cradle Switch Assembly
15	Bracket, Dial (L.H.)
16	Bracket, Dial (R.H.)
17	Bracket, Terminal Board
18	Spacer
19	Printed Circuit Network
20	Card, Number
21	Retainer, Number Card
22	Housing Assembly
23	Housing
24	Screw
25	4200-opg Tone Dial Assembly
26	Cord, Modular (Handset)

THE 200 TELEPHONE

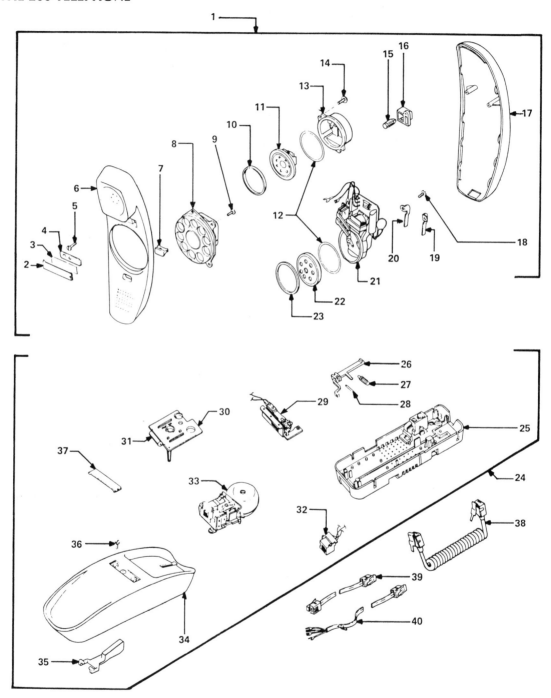

Fig. C-32 *Exploded view of the 200*
Trendline telephone (Courtesy of Cortelco/ITT Phone Systems).

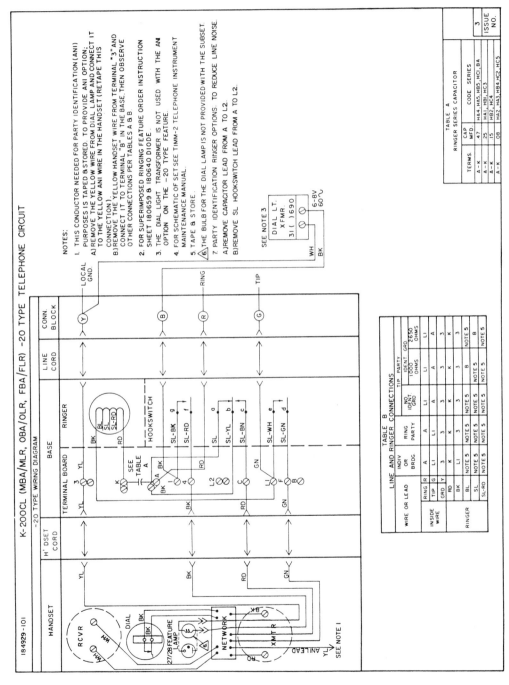

Fig. C-33 Schematic diagram of the rotary Trendline telephone (Courtesy of Cortelco/ITT Phone Systems).

Table C-11. Parts list for the 200 Trendline telephone.

Parts List for the 200 Telephone	
INDEX NUMBER	DESCRIPTION
1	Handset Assembly
2	Retainer, Number Card
3	Card, Number
4	Shield, Light
5	Screw, Housing
6	Housing
7	Button, Recall
8	1000-oog Tone Dial Assembly
9	Screw, Dial Mounting
10	Gasket, Receiver
11	Receiver
12	Gasket
13	Cup Assembly, Receiver
14	Screw, Cup Assembly Mounting
15	Lamp
16	Housing, Lamp
17	Cover, Handset
18	Screw
19	Bracket, Retainer (L.H.)
20	Bracket, Retainer (R.H.)
21	Modular Network Assembly
22	Transmitter
23	Gasket, Transmitter
24	Base and Housing Assembly (Standard) (Full Modular)
25	Base Assembly
26	Lever Arm
27	Spring, Lever Arm
28	Pin, Pivot
29	Cradle Switch Assembly
30	Terminal Board
31	Capacitor
32	Modular Jack
33	153-ABA Ringer
34	Housing
35	Plunger
36	Screw, Housing
37	Nameplate
38	Cord, Coiled
39	Cord, Line (Full Modular)
40	Cord, Line (Standard Spade Lug)

THE 2200/2300 TELEPHONE

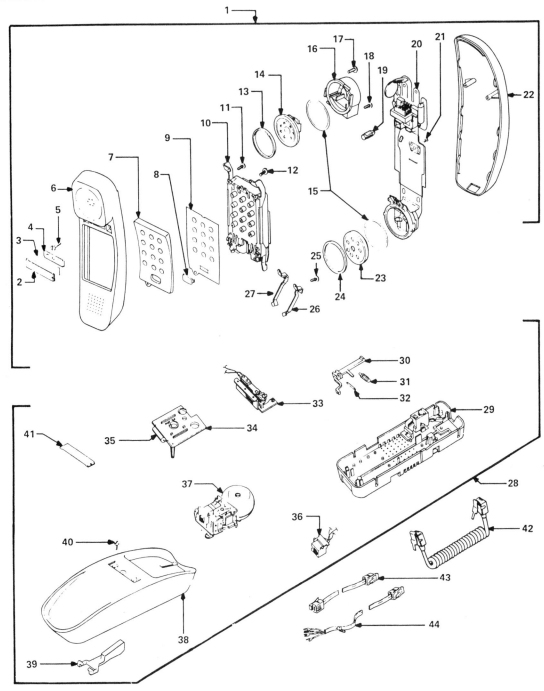

Fig. C-34 *Exploded view of the 2200/2300 telephone (Courtesy of Cortelco/ITT Phone Systems).*

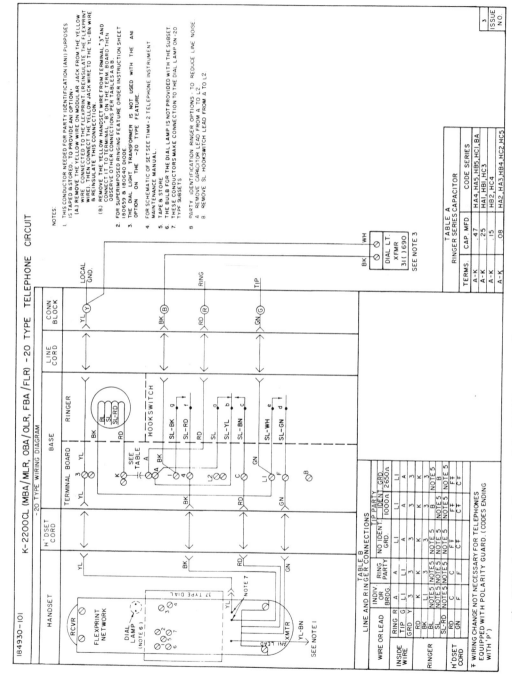

Fig. C-35 Schematic diagram of the 2200/2300 Trendline telephone (Courtesy of Cortelco/ITT Phone Systems).

Table C-12. Parts list for the 2200/2300 telephone.

Parts List for the 2200 Telephone	
INDEX NUMBER	DESCRIPTION
1	Handset Assembly
2	Retainer, Number Card
3	Card, Number
4	Shield, Light
5	Screw, Housing
6	Housing
7	Guide, Light
8	Button, Recall
9	Card, Dial
10	3700-000 Dial Assembly
11	Screw, Dial Mounting (Upper)
12	Screw, Dial Mounting (Lower)
13	Gasket, Receiver
14	Receiver
15	Gasket
16	Cup Assembly, Receiver
17	Screw, Cup Assembly Mounting
18	Screw, Flexprint Mounting
19	Lamp
20	Network Assembly (Flexprint)
21	Screw, Flexprint Mounting
22	Cover, Handset
23	Transmitter
24	Gasket, Transmitter
25	Screw
26	Bracket, Retainer (L.H.)
27	Bracket, Retainer (R.H.)
28	Base and Housing Assembly (Standard)
29	Base Assembly, Plastic
30	Lever Arm
31	Spring, Lever Arm
32	Pin, Pivot
33	Cradle Switch Assembly
34	Terminal Board
35	Capacitor
36	Modular Jack
37	153-ABA Ringer
38	Housing
39	Plunger
40	Screw, Housing
41	Nameplate
42	Cord, Coiled
43	Cord, Line (Full Modular)
44	Cord, Line (Standard Spade Lug)

☎D
Suppliers' Names and Addresses

WHETHER YOU ARE INTERESTED IN BUILDING A DIGITAL TELEPHONE LOCK, THE electronic telephone ringer, or the talking telephone, the first order of business is to locate a supplier of the needed electronic parts.

This might sound easier than it really is. Most likely, all components needed for any project will not be found at any one location. For this reason, you should request component catalogs from a number of distributors. To help you select and purchase electronic components, an extensive listing of distinguished mail order businesses that provide to the hobbyist first-quality electronic components at very competitive prices is contained in this appendix.

Request catalogs, flyers, or product listings from these fine companies. Then, when the information is received, it is up to you to make the proper choice of component type and value so that the circuits you design will operate the first time that you apply power. (Boldface indicates that these manufacturers were referred to in the text.)

MAIL ORDER ELECTRONIC COMPONENT DISTRIBUTORS

Active Electronics
P.O. Box 546 Highland Dr.
Westborough, MA 01581

Advanced Computer Products, Inc.
P.O. Box 17329
Irvine, CA 92713

All Electronics Corp.
P.O. Box 567
Van Nuys, CA 91408

Allied Electronics
401 E 8th St.
Fort Worth, TX 76102

American Design Components
62 Joseph St.
Moonachie, NJ 07074

Amp Supply Co.
73 Maple Dr.
Hudson, OH 44236

Apex Microtechnology Corp.
5980 N. Shannon Rd.
Tucson, AZ 85741

Atlantic Surplus Sales
3730 Nautilus Ave.
Brooklyn, NY 11224

ATV Research
13th & Broadway
Dakota City, NE 68731

BG Micro
P.O. Box 280298
Dallas, TX 75228

Babylon Electronics
P.O. Box 41778
Sacramento, CA 95841

Circuit Specialists
P.O. Box 3047
Scottsdale, AZ 85257

Del-Phone Industries
P.O. Box 5835
Spring Hill, FL 34606
(Request SASE for catalog)

Digi-Key Corp.
Highway 32 So. Box 667
Thief River Falls, MN 56701

Digital Research
P.O. Box 401247
Garland, TX 75040

Edmund Scientific
4128 Edscorp Building
Barrington, NJ 08007

Electronic Marketplace
5344 Jackson Dr.
La Masa, CA 92041

Fair Radio Sales Co.
1016 E Eureka St.
Lima, OH 45802

Fugi-Svea
P.O. Box 3375
Torrance, CA 90510

G&C Communications
Box 5632
1529 N Cotner Blvd.
Lincoln, NE 68505

Information Unlimited
P.O. Box 716
Amherst, NH 03031

International Crystal
10 N Lee
Oklahoma City, OK 73102

ITT Microsystems
Hillsboro Plaza
700 NW 12th Ave.
Deerfield Beach, FL 33441

Jameco Electronics
1355 Shoreway Rd.
Belmont, CA 94002

Jan Crystals
P.O. Box 06017
Fort Myers, FL 33906

JDR Microdevices
110 Knowles Dr.
Los Gatos, CA 95030

LZR Electronics Inc.
8174 Beechcraft Ave.
Gaithersburg, MD 20879

Mark V Electronics
8019 E Slauson Ave.
Montebello, CA 90640

Micro Mart
508 Central Ave.
Westfield, NJ 07090

Mouser Electronics
11433 Woodside Ave.
Santee, CA 92071

Newark Electronics
228 East Lake St.
Chicago, IL 60640

Parts Express
340 East First St.
Dayton, OH 45402

Quest Electronics
P.O. Box 4430
Santa Clara, CA 95054

Radio Shack
(Local malls and shopping centers)

Sintec Electronics
Drawer Q
Milford, NJ 088048

Solid State Sales
P.O. Box 74D
Somerville, MA 02143

Steven Mail Order Electronics
P.O. Box 698
Melville, NY 11747

MANUFACTURERS' NAMES AND ADDRESSES

In case you require additional information on the specialized integrated circuit used in many of the telephone projects in this book, an extensive listing of major semiconductor manufacturers is included.

The companies listed are well-known manufacturers of semiconductor devices. Technical information on any of their semiconductors devices as well as application notes and manuals, are available just for the asking.

Contact any one of the listed manufacturers when you need technical information on any integrated circuit or other semiconductor device.

Action Instruments
8601 Aero Dr.
San Diego, CA 92123

Advanced Analog
2270 Martin Ave.
Santa Clara, CA 95050

Advanced Micro Devices
901 Thompson Place
Sunnyvale, CA 94088

Advent Products
965 North Main St.
Orange, CA 92667

American Automation
14731 Franklin Ave.
Tustin, CA 92680

Amperex Electronic Corp.
Providence Pike
Slatersville, RI 02876

Analog Devices
One Technology Way
Norwood, MS 02062

Analogic Corp.
8 Centennial Dr.
Peabody, MA 01961

Analog Systems
P.O. Box 35879
Tucson, AZ 85740

Apex Microtechnology
1130 East Pennsylvania St.
Tucson, AZ 85714

Applied Micro Circuits
8808 Balboa Ave.
San Diego, CA 92123

Applied Microsystems
P.O. Box C-1002
Redmond, WA 98052

Applied Micro Technology
P.O. Box 3042
Tucson, AZ 85702

Applied Systems
26401 Harper Ave.
St. Clair Shores, MI 48081

Aptek Microsystems
Hillsboro Plaza
700 NW 12th Ave.
Deerfield Beach, FL 33441

Array Technology
922 South Saratago/Sunnyvale Rd.
San Jose, CA 95129

AWI
3212 Scott Blvd.
Santa Clara, CA 95050

Barvon Research Inc.
2680 N First St. Suite 210
San Jose, CA 95134

Beckman Instruments
2500 Harbor Blvd.
Fullerton, CA 92634

Bedford Computer Systems
4 Lyberty Way
Westford, MA 01886

Booktree Corp.
9950 Barnes Canyon Rd.
San Diego, CA 92121

Burr-Brown
P.O. Box 11400
Tucson, AZ 85734

CAE Systems
1333 Bordeaux Ave.
Sunnyvale, CA 94086

California Micro Devices
2000 W 14th St.
Tempe, AZ 85281

Cermetek Microelectronics
1308 Borregas Ave.
Sunnyvale, CA 94088

CGRS Inc.
P.O. Box 102
Langhorn, PA 19047

Cherry Semiconductor Corp.
2000 South County Trail
East Greenwich, RI 02818

Circuit Technology
160 Smith St.
Farmingdale, NY 11735

Citel Inc.
3060 Raymond St.
Santa Clara, CA 95050

Comark Corp.
93 West St.
Box 474
Medfield, MA 02052

Comlinear Corp.
2468 East 9th St.
Loveland, CO 80537

Commodore Semiconductor Group
950 Rittenhouse Rd.
Morristown, PA 19403

Computer Automation
18651 Von Karman
Irvine, CA 92713

Control Logic
9 Tech Circle
Natick, MA 01760

Creative Micro Systems
3822 Cerritos Ave.
Los Alamitos, CA 90720

Cromemco
280 Bernardo Ave.
Mountain View, CA 94039

Crystal Semiconductor
4210 South Industrial Dr.
Austin, TX 78744

Cubit Inc.
190 S. Whisman Rd.
Mountain View, CA 94041

Curtis Electro Devices
P.O. Box 4090
Mountain View, CA 94040

Custom Integrate Circuits
5353 Wayzata Blvd., Suite 603
Minneapolis, MN 55416

Custom MOS Arrays
211 Topaz St.
Milpitas, CA 95035

Cybernetic Micro Systems
P.O. Box 3000
San Gregorio, CA 94074

Cybersystems Inc.
7540-A South Memorial Parkway
Huntsville, AL 35802

Cynernetic Micro Systems
P.O. Box 3000
San Gregoria, CA 94074

Cypress Semiconductor
3901 N First St.
San Jose, CA 95134

Dallas Semiconductor
4350 S. Beltwood Parkway
Dallas, TX 75244

Data General Corp.
4400 Computer Dr.
Westborough, MA 01581

Data I/O
10525 Willows Rd. NE C-46
Redmond, WA 98052

Data Translation
100 Locke Dr.
Marlboro, MA 01752

Datel
11 Cabot Blvd.
Mansfield, MA 02048

Dense-Pac Microsystems Inc.
7321 Lincoln Way
Garden Grove, CA 92641-1428

Digelec Inc.
7335 East Acoma Dr., Suite 103
Scottsdale, AZ 85260

Digital Equipment
77 Reed Rd.
Hudson, MA 01749

Digital Microsystems
1755 Embarcaders
Oakland, CA 94606

Digitek
17505 68th NE
Box 468
Kenmore, WA 98028

Dionics Inc.
65 Rushmore St.
Westbury, NY 11590

Distributed Computer Systems
223 Crescent St.
Waltham, MA 02154

Diversified Technology
Box 465
112 East State St.
Ridgeland, MI 39157

Dumont Alphatron
10351 Bubb Rd.
Cupertino, CA 95014

EG&G Reticon Corp.
343 Potrero Ave.
Sunnyvale, CA 94086

E-H Electronics
7303 Edgewater Dr.
Oakland, CA 94621

Electronic Designs Inc.
42 South St.
Hopinton, MA 01748

EMM-SESCO
20630 Plummer St.
Box 668
Chatsworth, CA 91311

Emulogic
3 Technology Way
Norwood, MA 02062

ETI Corp.
6918 Sierra Ct.
Dublin, CA 94568

EXAR Integrated Systems
2222 Qume Dr.
San Jose, CA 95161

EXEL Microelectronics Inc.
500 Valley Way
Milpitas, CA 95035

Fairchild
10400 Ridgeview Ct.
Box 1500
Cupertino, CA 95014

Ferranti
87 Modular Ave.
Commack, NY 11725

Farranti/Interdesign
1500 Green Hills Rd.
Scotts Valley, CA 95066

Fujitsu America
910 Sherwood, Suite 23
Lake Bluff, IL 60044

Fujitsu Microelectronics Inc.
3545 North First St.
San Jose, CA 95134

GE Solid State Inc.
P.O. Box 2900
Somerville, NJ 08876

General Instrument
600 W. John St.
Hicksville, NY 11802

General Micro Systems
1320 Chaffey Ct.
Ontario, CA 91762

Gennum Corp.
P.O. Box 489, Station A
Burlington, Ontario
Canada L7R 3Y3

Goldstar Semiconductor Ltd.
1130 E. Arquez Ave.
Sunnyvale, CA 94086

Gould Semiconductors
2300 Buckskin Rd.
Pocatello, ID 83201

GTE
2000 West 14th St.
Tempe, AZ 85281

Harris Semiconductor
P.O. Box 883
Melbourne, FL 32901

Harris Digital Products Div.
724 Route 202
Somerville, NJ 08876

Heurikon
3001 Latham Dr.
Madison, WI 53713

Hewlett-Packard
11000 Wolfe Dr.
Cupertino, CA 95014

Hitachi America
2210 O'Toole Ave.
San Jose, CA 95131

Holt Inc.
8 Chrysler St.
Irvine, CA 92714

Honeywell
1150 E. Cheyenne Mountain Blvd.
Colorado Springs, CO 80906

Hughes
500 Superior Ave.
Newport Beach, CA 92658

Hybrid Systems
22 Linnell Circle
Billerica, MA 01821

Hy Comp Inc.
75 Union Ave.
Box 377
Sudbury, MA 01776

Hyundai Elexs America
166 Baypointe Parkway
San Jose, CA 95134

ILC Data Device
105 Wilbur Place
Bohemia, NY 11716

Industrial Micro-Systems
189 Hitchcock Rd.
Southington, CO 06489

Infosphere Inc.
4730 SW Macadam Ave.
Portland, OR 97201

Inmos Corp.
1110 Bayfield Rd.
Colorado Springs, CO 80906

Inmos
P.O. Box 16000
Colorado Springs, CO 80935

Intech-Advanced Analog
2270 Martin Ave.
Santa Clara, CA 95050

Interlink Electronics
535 E. Montecito
Santa Barbara, CA 93103

Integrated Circuit Engineering
15022 N 75th St.
Scottsdale, AZ 85260

Integrated Circuit Systems
1012 W 9th Ave.
King of Prussia, PA 19406

Integrated Device Technology
3236 Scott Blvd.
Box 58015
Santa Clara, CA 95052 – 8015

Integrated Technology
1233 N Stadem Dr.
Tempe, AZ 85281

Intel
3065 Bowers Ave.
Box 58065
Santa Clara, CA 95051

Interdesign
1500 Green Hills Rd.
Scotts Valley, CA 95066

Intergraph, Advanced Processor Div
2400 Geng Road
Palo Alto, CA 94393

International Cybernetics
2270 Westbury Ave.
Clearwater, FL 33516

International Microcircuits
3350 Scott Blvd. Bldg. 37
Santa Clara, CA 95051

International Microelectronics
  Products
2830 N 1st St.
San Jose, CA 95134

International Microsystems
1154 C Ave.
Auburn, CA 95603

Intersil
2450 Walsh Ave.
Santa Clara, CA 95051

Intronics
57 Chapel St.
Newton, MA 02158

ISL Corp.
4354 Olive St.
St. Louis, MO 63108

ITT Semiconductors
7 Lake St.
Lawrence, MA 01841

Kinetic Systems
11 Maryknoll Dr.
Lockport, IL 60441

Kontron Electronics
630 Price Ave.
Redwood City, CA 94063

Lambda Semiconductors
121 International Dr.
Corpus Christi, TX 78410

Linear Technology Corp.
1630 McCarthy Blvd.
Milpitas, CA 95035-7487

LSI Computer Systems
1235 Walt Whitman Rd.
Melville, NY 11747

LSI Logic Corp.
1601 McCarthy Blvd.
Milpitas, CA 95035

Linear Technology
1630 McCarthy Blvd.
Milpitas, CA 95035

3M Products
P.O. Box 2963
Austin, TX 78769

Master Logic
761 E Evelyn Ave.
Sunnyvale, CA 94086

Matrix Corp.
1639 Green St.
Raleigh, NC 27603

Matrox Elec. Systems
5800 Andover Ave. T.M.R.
Montreal, Quebec H4T 1H4, Canada

Maxim
510 Pastoria Ave.
Sunnyvale, CA 94086

Micrel
1235 Midas Way
Sunnyvale, CA 94086

MIPS Computer Systems
928 Arques Ave.
Sunnyvale, CA 94086

MCE
1111 Fairfield Dr.
W. Palm Beach, FL 33407

Microcircuits Technology
1157 San Antonio Rd.
Mountain View, CA 94043

Micro Computer Control
P.O. Box 275
Hopewell, NJ 08525

Micro Innovators
2348A Walsh Ave.
Santa Clara, CA 95051

Micro-Link Corp.
14602 N US Hwy. 31
Carmel, IN 46032

Micron Technology Inc.
2805 E Columbia Rd.
Boise, ID 83706

Micro Networks
324 Clark St.
Worcester, MA 01606

Micropac Industries
905 E Walnut St.
Garland, TX 75040

Micro Power Systems
3151 Jay St.
Box 54965
Santa Clara, CA 95054

Micro Sciences
145 Commank Rd.
Commack, NY 11725

MilerTronics
303 Airport Rd.
Greenville, SC 29607

Miller Technology
647 N Santa Cruz Ave.
Los Gatos, CA 95030

Mitel Semiconductor
P.O. Box 13320, Kanata
Ontario, Canada K2K 1X5

Mitsubishi
1050 E. Arques Ave.
Sunnyvale, CA 94086

Monolithic Memories
2175 Mission College Blvd.
Santa Clara, CA 95054

Monolithic Systems
84 Inverness Circle E
Englewood, CO 80112

Monosil
3060 Raymond St.
Santa Clara, CA 95050

Motorola
6501 William Cannon Dr. W
Austin, TX 78735

MX-Com
4800 Bethania Station Rd.
Winston-Salem, NC 27105

National Semiconductor
2900 Semiconductor Dr.
Santa Clara, CA 95051

NCM Corp.
1500 Wyatt Dr., Suite 9
Santa Clara, CA 95054

NCR
8181 Byers Rd.
Miamisburg, OH 45342

NEC Electronics Inc.
401 Ellis St.
Mountain View, CA 94039

Nitron Inc.
10420 Bubb Rd.
Cupertino, CA 95014

Octagon Systems
6501 West 91st Ave.
Westminster, CO 80030

Ohio Scientific
1333 S Chillcothe Rd.
Aurora, OH 44202

OKI Semiconductor Inc.
650 N. Mary Ave.
Sunnyvale, CA 94086

Oliver Advanced Engineering
676 W Wilson Ave.
Glendale, CA 91203

Omnibyte Corp.
245 W Roosevelt Rd. Bldg. 1-5
West Chicago, IL 60185

Onset Computer Corp.
199 Main St.
Box 1016
North Falmount, MA 02556

Optical Electronics
P.O. Box 11140
Tucson, AZ 85734

Panasonic Industrial
425 E Algonquin Rd.
Arlington Heights, IL 60005

Pico Design
1333 Lawrence Expressway,
   Suite 340
Santa Clara, CA 95051

Plessey Solid State
1500 Green Hills Rd.
Scotts Valley, CA 95056

Polycore Electronics
1107 Tourmaline Dr.
Newbury Park, CA 91320

Precision Monolithics
1500 Space Park Dr.
Santa Clara, CA 95052

Pro-Log
2411 Garden Rd.
Monterey, CA 93940

Quay
P.O. Box 783
Eatontown, NJ 07724

Raytheon Semiconductor
350 Ellis St.
Mountain View, CA 94039

RCA
Route 202
Somerville, NJ 08876

RCI/Data
1992 Lakewood Rd.
Toms River, NJ 08753

Relms Memory Systems
1180 Miraloma Way
Sunnyvale, CA 94086

Recticon
345 Potero Ave.
Sunnyvale, CA 94086

Riehi Time Corp.
53 S Jefferson Rd.
Whippany, NJ 07981

RIFA Inc.
403 International Pkwy
Richardson, TX 75085 – 3904

Rockwell International
4311 Jamboree Rd.
Box C
Newport Beach, CA 92658

SMOS Systems
50 West Brokaw Rd., Suite 7
San Jose, CA 95110

Samsung Semiconductors
3725 North First St.
Santa Clara, CA 95134

Sanyo Semiconductors
7 Pearl Ct.
Allendale, NJ 07401

SBE Inc.
4700 San Pablo Ave.
Emeryville, CA 94608

SEEQ Technology Inc.
1849 Fortune Dr.
San Jose, CA 95131

Semi Processes
1971 N Fortune Dr.
San Jose, CA 95131

SGS Semiconductor
1000 E Bell Rd.
Phoenix, AZ 85022

Sharp Electronics
10 Sharp Plaza
Paramus, NJ 07652

Siemans
2191 Laurelwood Rd.
Santa Clara, CA 95054

Si-Fab Corp.
27 Janis Way
Scotts Valley, CA 95066

Signetics
811 E Arques Ave.
Sunnyvale, CA 94088

Silicon General
11861 Western Ave.
Garden Grove, CA 92641

Silicon Systems
14351 Myford Rd.
Tustin, CA 92680

Siliconix Inc.
2201 Laurelwood Rd.
Santa Clara, CA 95054

Siltronics Ltd.
436 Highway 7
Kanata, Ontario K2L 1T9 Canada

Solarise Enterprises
10080 N Wolfe Rd., Suite SW3-180
Cupertino, CA 95014

S-MOS
2460 North First St.
San Jose, CA 95131

Solid State Micro Technology for
  Music Inc.
2076B Walsh Ave.
Santa Clara, CA 95050

Solitron Devices Inc.
1177 Blue Heron Blvd.
Riveria Beach, FL 33404

Sprague Solid State
3900 Welsh Rd.
Willow Grove, PA 19090

Stancor
131 Godfrey St.
Logansport, IN 46947

Standard Microsystems
35 Marcus Blvd.
Hauppague, NY 11788

STD Microsystems
399 Sherman Ave.
Palo Alto, CA 94306

STS Thomas Micro Electronics
1310 Electronics Dr.
Carrollton, TX 75006

Stynetic Systems
Flowerfield—Bldg. 1
Saint James, NY 11780

Tektronix
P.O. Box 500
Beaverton, OR 97077

Teledyne Crystalonics
147 Sherman St.
Cambridge, MA 02140

Telephonics LSI
790 Park Ave.
Huntington, NY 11743

Telmos
740 Kiler Rd.
Sunnyvale, CA 94086

Teltone Corp.
10801-120th Ave. NE
Kirkland, WA 98033

Texas Instruments
P.O. Box 401560
Dallas, TX 75240

Thomson Components Corp.
6203 Variel Ave.
Woodland Hills, CA 91367

Toshiba America Inc.
2692 Dow Ave.
Tustin, CA 92680

TRW
P.O. Box 2472
La Jolla, CA 92038

United Microelectronics Corp.
1575 Garden of the Gods Rd.
Colorado Springs, CO 80907

Universal Semiconductor
1925 Zanker Rd.
San Jose, CA 95112

Vitesse Semiconductor
741 Calle Plano
Camarillo, CA 93010

VLSI Technology
1101 McKay Dr.
San Jose, CA 95131

Votrax
500 Stephenson Highway
Troy, MI 48084

Weitek
1060 E Arques Ave.
Sunnyvale, CA 94086

Wintek Corp.
1801 South St.
Layfayette, IN 47904

Xicor
851 Buckeye Ct.
Milpitas, CA 95035

Xycom Corp.
750 N Maple Rd.
Saline, MI 48176

Zilog Inc.
210 Hacienda Ave.
Cambell, CA 95008

Zytrex Corp.
224 North Wolfe Rd.
Sunnyvale, CA 94086

ZyMOS Inc.
P.O. Box 62379
Sunnyvale, CA 94088

FREE MAGAZINES

Following is a listing of free magazines, called *trade journals*. To get a subscription to any or all magazines listed, you must meet eligibility requirements. One requirement is that you have your own business in either the electronics or computer fields. If you met the minimum requirements, just drop the publishers a note on your company letterhead. They will be more than happy to send you a questionnaire that you must fill out and return.

Business Electronics
Willow Publishing Co.
464 Central Ave.
Northfield, IL 60093

Computer Design
One Technology Park Drive
P.O. Box 990
Westford, MA 01886

Electronic Business
The Cahners Publishing Company
275 Washington St.
Newton, MA 02158

Electronics Test
Miller Freeman Publications Inc.
500 Howard St.
San Francisco, CA 94105

Electronic Products
Hearst Business Communications
   Inc.
645 Stewart Ave.
Garden City, NY 11530

Manufacturing Systems
Hitchcock Publishing Co.
P.O. Box 3008
Wheaton, IL 60189-9972

Personal Engineering
Box 1821
Borrkline, MA 02146

Solid State Technology
1421 South Sheridan
P.O. Box 3689
Tulsa, OK 74101 – 3689

The listing of companies is not a product endorsement of any kind. The sole purpose of this listing is to enlighten the hobbyist or technician to the names and addresses of potential suppliers of computer and electronic goods as well as integrated circuits. It is up to readers to determine which company suits their individual needs.

*DREAMS OF YESTERDAY
IS THE HOPE OF TODAY
AND THE REALITY OF TOMORROW*

Robert Goddard
Father of Modern Rocketry

Index